Identified Flying Objects

A Multidisciplinary Scientific Approach to the UFO Phenomenon

Identified Flying Objects

A Multidisciplinary Scientific Approach to the UFO Phenomenon

Dr. Michael P. Masters

Printed in the United States of America

P.O. Box 461

Butte, MT 59703-0461

Ordering Information:

Special discounts are available for quantity purchases.

Order requests by U.S. trade bookstores and wholesalers should be sent to the author via the contact page at:

https://idflyobj.com or in writing at the above address.

Cover Design: Michael Masters

Cover Images: Ascent of man ending with smartphone - Frank Fiedler/shutterstock.com; Extratempestrial modified from silhouette of modern human created by Anna Rassadnikova/shutterstock.com; UFO center image by Gl0ck/shutterstock.com

First Edition – ISBN: 978-1-7336340-6-9

Printed in the United States of America

This book is dedicated to my patient and loving family, friends, sunshine and smiles.

Dr. Michael P. Masters is a professor of Biological Anthropology at Montana Tech in Butte, Montana. After receiving a Ph.D. in Anthropology from The Ohio State University in 2009, where he specialized in hominin evolutionary anatomy, modern human variation, archaeology, and biomedicine, Dr. Masters spent the following decade developing a broad academic background that unites multiple scientific disciplines with the aim of elucidating a currently unexplained phenomenon.

Remaining vigilant in his own skepticism, Dr. Masters continues this research with the intent of initiating informed dialog about UFOs via an abductive method of inquiry that is firmly rooted in science and the principle of parsimony. Collectively, Dr. Masters' background, education, and current research program combine to offer a unique perspective and a novel approach to addressing unanswered questions pertaining to a widely recognized, yet poorly understood aspect of modern global culture.

Website:

www.idflyobj.com

Twitter:

@MorphoTime

Facebook:

@idflyobj

In any field, find the strangest thing and then explore it.[i]

− John Archibald Wheeler

Table of Contents

Acknowledgements

I would like to offer sincere gratitude to the following individuals for their contribution to the evolution of this multi-year research project, including Vincent Siragusa, Heidi Reid, Jason VonVille, Arron Carnahan, Daniel Drouin, Valerie Moring Shubert, Erica Jansma, Annika Rapp, Cheyenne Crooker, BreAnna Wright, Matt Boyle, and a special thanks to the peer reviewers, whose comments and suggestions helped immensely in improving the content, clarity, and flow of the manuscript.

1

Shifting the Debate

All we can ask of a theory is to predict the results of events that can be measured. [2]

— Leon M. Lederman

An idea that is not dangerous is unworthy of being called an idea at all. [3]

— Oscar Wilde

1.1. Conception

It is often difficult to separate fact from fiction, and particularly at a young age. Though beginning with the onset of object permanence, and throughout the remainder of infancy and childhood, we begin to develop a sense of what is and is not a part of our natural world. Through the simple act of living, we are constantly evolving a perception of reality that both reflects and creates the cultural norms of our society. By the age of eight years old, I was confident that I had become rather adept at deciphering fact from fiction, or at least until learning of an odd encounter my father once had with an Unidentified Flying Object (UFO), some years earlier, in the dark skies above Amish country in rural Ohio.

As a veterinarian, my dad often responded to late-night farm calls in relatively remote parts of northeast Ohio. On one of these nights, while driving to a call with a colleague who happened to be riding along with him, he approached the crest of a hill and noticed a bright light in the distance. I remember him describing it as a glowing ball of light sitting near the horizon, just over the next hillside. However, unlike most lights that emanate beams outward from the center, this object just glowed brightly like the full moon on a clear summer night, while hovering silently off in the distance.

It was patently clear to him that this was not the moon, a star, a weather balloon, a streetlight, or the headlights of an oncoming car, particularly considering how uncommon the latter three of these are in the heart of Amish country. Further confirmation that this was not an ordinary occurrence came when this glowing ball of light suddenly darted toward their truck and stopped only a couple hundred meters away. This strange light hovered there for a brief moment, rapidly zipped back across the horizon away from them, stopped again, and then shot upward at tremendous speed as it disappeared from sight.

Naturally, this sort of occurrence can be difficult to process and can illicit strong feelings of wonder, excitement, and fear. In fact, my mother vividly remembers sensing all three of those emotions as my father relayed the events of that night back to her over the CB radio. I too remember being filled with some degree of astonishment as my parents told this story to some friends at our house while I eavesdropped from the stairs, long after I was meant to be in bed. After hearing this account for the first time, I worked to make sense of this new and rather dubious facet of my previously established sense of what was real and what was not. Although I had certainly heard of UFOs, they were always presented in the context of science fiction, and never as something witnessed by real people who I knew and trusted.

Not long after this encounter, my father bought a book entitled *Communion*, which details the author's own experiences with UFOs and the human-like beings who purportedly pilot them.[4] Although it would be decades before I actually got around to opening it, the book's cover proved to be somewhat influential in shaping my perception of this phenomenon. In fact, even some thirty years later, I can still vividly recall the moment I looked up and saw that book on the living room shelf, as well as the oddly specific image that entered my mind at that time. It was a rather simple and fleeting mental image, which encompassed three separate forms visualized together. On the left side was something akin to a chimpanzee, in the middle was a modern human form, and on the right side was that odd, yet entirely familiar humanoid alien creature from the book cover, with an enlarged hairless cranium, big eyes, a small nose and mouth, thin lips, and a narrow chin (figure 1-1).[5]

Figure 1-1: *Rudimentary approximation of this edifying mental image as seen at age eight.*

As a young child growing up in a small town in the honorary bible belt of the American Midwest, I knew very little about the process of biological change, and even less about the long history of human evolution on this planet. Though even then, it was evident that the human in the center of this mental portrait resembled both the "chimpanzee" to the left, and the "alien" to the right, which led me to wonder if they could somehow be related.

Over time, I came to understand in great detail the process of evolutionary change and the phylogenetic relationships among extant (living) and extinct (nonliving) organisms. While developing this scholastic knowledge, it became increasingly clear how each of these three seemingly distinct forms could all come to possess such similar traits—or synapomorphies, as I would later understand them—if they all shared the same common ancestor as part of a shared evolutionary history here on Earth.

Furthermore, if the suite of characteristics common to each of these three forms was a result of common ancestry, then these "extraterrestrial" aliens would not be extraterrestrial at all. Rather, they may be better understood as the product of further human evolution on this planet, following many millennia of continued biological and cultural change. In other words, these "aliens"—rather than being from a different planet in a distant solar system—may simply be us, from a different time in the distant future.

1.2. Ontogeny

Our tiny branch within the 3.5-billion-year-old tree of life is known as the hominin lineage. It began about 6-8 million years ago, when our ancestors started to diverge from what would eventually become the common (*Pan troglodytes*) and pygmy (*Pan paniscus*) chimpanzees, our closest living relatives on this planet. Although we each possess a number of unique biological and behavioral traits, what most clearly defines the human lineage is our bipedal form of locomotion (i.e., habitual upright walking). This seemingly insignificant change in the way we get from one place to another actually had far-reaching implications for the subsequent development of our advanced culture and biology.

Not least among these was the impact that bipedalism had on our brain size, as it helped to create more space within the skull for a larger brain to grow. Standing and walking upright also freed our hands from the burdens of moving our bodies, which meant that our hands and highly dexterous digits could now be used for all sorts of new and innovative tasks, which also helped to advance our culture and intellect. In the context of the aforementioned mental picture (partially invoked by the cover image on the book *Communion*), these shared traits alone suggested to my primitive 8-year-old mind that modern humans could potentially represent an intermediate stage between this small-brained/large-faced chimpanzee-like (early hominin) form, and the large-brained/small-faced alien-like (extraterrestrial) form.

As I worked toward becoming a professor of biological anthropology, with a specialization in human evolutionary anatomy and modern human variation, it became increasingly evident that if these "aliens" are in fact real, then they must be part of the hominin lineage, and clearly toward the future end of our current position along the fourth dimension of space-time. Additionally, in the same way that we have come to understand the current human condition by examining the morphology and culture of our hominin ancestors, it seemed reasonable that these "aliens of time" may also be working to probe their own evolutionary past, by dint of the much more sophisticated anthropological tool of time travel. After all, what anthropology instructor hasn't at some point uttered "if only we had a time machine."

It is important to acknowledge—particularly given the young age at which this notion arose—that my conviction regarding this time travel interpretation of the UFO phenomenon could have been strengthened over time if I had only been looking for evidence that supported it. This form of intellectual partiality is known as *confirmation bias*, which, stated more specifically, is when people develop a propensity to interpret things in a way that supports conclusions they have already drawn about something.

Confirmation bias and other sources of bias should always be considered by researchers, as well as by those reading the results of research conducted by others. Though even with a keen awareness of this and other logical fallacies—as well as a persistently critical perception of the idea as a whole—I came across very few lines of evidence that raised doubts about this cross-temporal interpretation of the UFO phenomenon while conducting extensive research into the matter. To the contrary, the further I descended down the rabbit hole of interdisciplinary inquiry in an effort to investigate the possibility that our distant human descendants may someday reverse the flow of time to research us in their own past, the more plausible this incredible scenario became.

As a scientist trained to be incredulous in assessing intentions, methods, results, and the interpretation of results, as well as one who is fervently aware of longstanding manipulation of the scientific process in former and ongoing UFO inquests, I continue to remain vigilant in my own skepticism. I do not have a staunch unwavering devotion to this notion and I do not claim to proclaim a truth. Instead, my intent is simply to initiate dialog regarding this potentiality, by means of a method of inquiry that is firmly rooted in science and the principle of parsimony.

1.3. Science, Rhetoric, the Unknown, and the Unknowable

Recently, numerous pseudoscientific studies of UFOs and extraterrestrials have emerged, which have largely been based on conjecture and *speculocatenation*—the linking together of purely speculative ideas.[6] Unfortunately, this has acted to diminish the integrity of actual research into the matter and has significantly muddied the waters with regard to what constitutes an actual objective scientific investigation of the phenomenon. As Anne Cross stated in the 2004 article entitled *The Flexibility of Scientific Rhetoric: A Case Study of UFO Researchers*:

> Throughout its ups and downs, the development of the UFO research movement is a story, first of a lost battle for a place within the scientific establishment, forced exile from the mainstream scientific community, and, finally, the construction of a successful rhetorical and cultural strategy that uses science to garner legitimacy in the eyes of lay followers . . . Because Ufologists draw primarily on signifiers of science, rather than the substance of scientific knowledge and its methodologies, it appears that Ufologists' efforts are not directed at convincing mainstream scientists of their legitimacy. Instead, the strategy is directed at convincing laypersons.[7]

This article provides a valid critique of attempts at misrepresenting science in order to mislead the masses and offers notable examples of how scientific rhetoric has been stretched to accommodate a non-scientific pursuit of the unknown. However, in the context of the current hypothesis and the evidence put forth in support of it, these same criticisms are not entirely relevant, for a few important reasons.

First, the nature of the question posed in this book is different, as it is centered on actual tangible future outcomes, rather than simply speculating about whether something may or may not exist somewhere else in the vastness of space. Secondly, this book is written for academics as much as it is for those not directly involved in scientific research. As such, this text speaks directly to those who are most able to censure any flagrant attempt at using misleading terms, concepts, or even established knowledge, in an attempt to garner legitimacy among lay readers. Lastly, this book aims to maintain brazen honesty and persistent forthrightness regarding what can and cannot be known under the current proposed model, espousing a time-travel explanation for this phenomenon.

Rather than using thaumaturgy to convince the unsuspecting masses about things that cannot currently be known, this study instead draws from peer-reviewed research conducted by leading scholars in the fields of anthropology, astronomy, astrobiology, and physics, who are not involved in UFO research of any kind. As such, it is hoped that this inquiry can withstand intense scientific scrutiny by established members of each of these respective fields, as it is written for them as much as anyone else.

Furthermore, to counter the candid criticisms of Anne Cross (2004) with regard to misleading, unscientific, and non-academic approaches that have been used previously in UFO research, this manuscript underwent extensive pre-publication peer review by academics in fields related to the content of this book. More specifically, the manuscript was reviewed by a PhD in biology, a PhD in biological anthropology, and a PhD in theoretical and computational chemistry who has taught quantum mechanics, theoretical physics, and computational methods for over 20 years.

It should also be noted that I am not a Ufologist. Naturally, in order to write an informed book on the topic, some facets of UFOs and the sentient beings associated with them must be skeptically considered and discussed. However, this represents only a small part of what is otherwise a broad-based examination of long-term patterns of biological, cultural, and technological change throughout hominin evolution.

Moreover, it is important to emphasize that proof of the existence of UFOs and aliens, now or in the past, is not an essential prerequisite for this new model examining the phenomenon in the context of persistent long-term evolutionary changes in the hominin lineage. This is simply because the current hypothesis predicts that we will eventually become them and, as such, validation or refutation of this predicted state of our distant progeny will inevitably be revealed by the passage of time itself. In fact, this is a critical component of the current model, which distinguishes it from the prevailing paradigm of extraterrestrial life and interstellar space travel. More specifically, demonstrating the existence or non-existence of intelligent life elsewhere in the universe involves a much more tenuous search, with far more places to look than simply toward a future point in human time on this planet.

1.4. Temporal Bounds of Understanding

The most critical component of a scientific hypothesis is that it must be testable and falsifiable, and though it may seem counterintuitive, the current proposed time-travel model actually adheres to this requirement. More specifically, the continued existence (or extermination) of humanity on this planet innately allows this hypothesis to be tested and falsified. Fortunately, there is no statute of limitations on a working hypothesis. Rather, the longer an idea is around, the more opportunities there are to falsify it, which, while slow, is good for the scientific process. For as time passes without refutation and evidence in support of an idea mounts, the hypothesis grows and strengthens as it moves toward potentially becoming an accepted theory or law.

Take the process of evolutionary change for example, which is a key component of the current proposed model. Charles Darwin and Alfred Russel Wallace first proposed this novel and comprehensive explanation for the diversity of life on earth in the mid-1800s[8] and, while accepted as law by those who understand it (hence my not referring to it as evolutionary "theory"), there remain others who still attempt to refute it. Currently, the majority of challenges brought against evolution are the result of misapprehension, misinformation, bias, religious conviction, and perhaps even intellectual insecurity. However, although misguided, the questioning process itself is healthy, as it pushes researchers to develop and address novel inquiries in new and different ways.

The main objective of this ever-evolving theory of time travel and human evolution is to spur dialog regarding the holistic existence of our species. It is also to examine whether we may someday unravel the mysteries of time travel, and if we could already be seeing indications of this future outcome in the form of our distant descendants researching us in their own past. Additionally, this inquiry seeks to elucidate questions such as: what are UFOs, who is inside, why do they and their craft look the way they do, what are they doing, why are they so commonly reported doing the same types of things, what would be the motivation for doing these things, and what might all of this reveal about the future state of our species if they are indeed our distant human descendants?

However, without the advantage of future hindsight, and because we cannot see the many complex factors leading up to that future from which we may someday return, there are bound to be many more questions than answers. Even those who report having close contact with what are presumed to be our distant human progeny see only the end result of a long and multifaceted evolutionary process. While we may be able to infer some things from what they report seeing, it is impossible to know exactly what forces will contribute to that future human state between now and then, as it remains veiled by our obscure and enigmatic future.

Without the ability to become unstuck from time, we are limited by when we can be and how far we can see in either direction. As such, the purpose of this book is not to make predictions about specific future environments, or how the forces of evolution may shape our species in response to these disparate conditions. Instead, the majority of evidence regarding the probable future state of our distant human descendants is drawn from long-term morphological, cultural, and technological trends, which have occurred across vast stretches of our own much more per- ceptible past.

It would certainly be helpful to be able to draw from first-hand expe- riences with our anachronous scion. However, I have never had such an encounter, nor have I ever even seen a UFO that could not be explained by some other phenomenon. Subsequently, this smaller component of the current investigation must draw from the testaments of others, who describe in intricate detail a set of experiences that are exceedingly con- sistent, regardless of when or where they occurred throughout the world.

Archived reports provided by credible individuals and institutions are considered in association with the scientific evidence described throughout this text.[9 10 11 12] Such accounts are certainly not the primary focus of this inquiry. However, these resources are important to consider in conjunction with other lines of evidence, and may be of interest to individuals who have had their own encounter with a UFO, or to those who lend credence to the experiences of countless others.

Moreover, it is important that these individuals not be immediately dis- missed as crazy, drunk, high, psychotic, etc., as was far too often the case over the last 70 years. Instead, they and their distinctive narratives should be considered rationally and objectively, as they may represent an integral tool that could help provide a better understanding of this misconstrued phenomenon, as well as a broader conceptualization of humanity through deep time. After all, in addition to formal eyewitness accounts provided by numerous law enforcement officers and high-ranking military officials, even the former governor of Arizona, Fife Symington, staunchly asserts that he, along with thousands of other Arizonans, had observed a UFO in the skies above Phoenix in 1997.[13]

1.5. Looking Forward Toward Us Looking Back

Modern human cultural complexity is primarily the result of incremental changes that have occurred since we last shared a common ancestor with chimpanzees approximately 6 million years ago. We are fortunate to have at our disposal a wealth of skeletal and fossil specimens, as well as a large sample of tools and other materials left behind by our human ancestors. However, because we are stuck moving along a seemingly linear path through time, our view of the future remains shrouded in ambiguity. Nonetheless, if reports of close encounters with UFOs and "aliens" can be understood as instances of intertemporal interaction, they could potentially offer up a wealth of information about the future state of our species.

If time travel technology were currently available to anthropologists of our own time, there is no doubt that we would be using it to gain a much deeper understanding of our own past. For instance, we may choose to return to East Africa 1.5 million years ago to examine the biology and culture of *Homo erectus*. Additionally, if they were able to recognize us as their distant descendants, they would be offered the opportunity to learn something of their own future biology and culture, simply as a consequence of us investigating our own past.

In the same way that *Homo erectus* would struggle to recognize modern humans and the intricate instrumentation we currently possess, it seems natural that modern humans would also be limited in our ability to comprehend the morphology and technology of future humans investigating their own past. Though by examining the long-term biological and cultural trends that have gotten us to where we are today—while also taking into account what may be fleeting glimpses of our own future—the broader composite mosaic of human time may begin to come into view.

1.6. Cladistics, Classification, and Catalyzing Contention

Biological organisms change over time. The speed of this change varies depending on the rate of environmental modification, and the relative influence of the four forces of evolution: gene flow, gene drift, mutation, and natural selection. Examining shared characteristics among organisms that are also present in the ancestors of those species—known as *synapomorphies* or *shared derived characteristics*—allows us to identify evolutionary relationships among them.

If we can take into consideration reports provided by sound-minded individuals who assert that they have seen what are presumed to be our future human descendants, it is clear that both they and we share a number of derived characteristics that are unique to the hominin lineage. Among the most recognizable of these shared traits is bipedalism. In fact, this is a rare form of locomotion among all animals on Earth and, as will be discussed later in the context of astrobiology, is likely to be even rarer on earthlike exoplanets elsewhere in our known universe.

In addition to bipedalism, which is the trait that defines our hominin lineage above all others, reports of close encounters also suggest that we share bilateral symmetry, the lack of a tail, relative hairlessness, highly dexterous hands and fingers, a large and globular brain, large eyes, and small noses and mouths. Additionally, and perhaps most importantly, these synapomorphies could not exist if the alien creatures in these reports had undergone a separate evolutionary trajectory on another planet in a different solar system.

It is extremely unlikely that extraterrestrial life evolving on a different planet would ever develop traits so similar to those of our own species, genus, or even the whole of the primate order. It is also improbable that an intelligent lifeform on a distant planet would ever be able to locate us around one of the many billions of stars in our nondescript galaxy, or that we would possess the same level of technological advancement, and at the same time, so as to facilitate mutual discovery. Even if contact were made, would we be able to communicate with each other, or traverse thousands or millions of light-years of space to visit one another? Furthermore, if visitation were possible, extraterrestrials would not be expected to simply observe us from a distance or perform covert medical examinations on us, then simply return home without any formal contact.

Because an extraterrestrial explanation for the UFO phenomenon is so implausible, it seems compulsory to modify the language used to discuss it. For instance, "extraterrestrial" is the term used to refer to these alien creatures under the current space travel model. However, these pioneers of time are in all likelihood terrestrial, or earth-dwelling, just like us. In fact, we may well live, work, and play in the exact same space as our distant descendants, in the same way we now reside in the same space as those living on Earth 25,000 years ago, 500 years ago, or as recently as a few minutes or seconds ago. So, in order to stay consistent with this proposed paradigm shift concerning the origin of these alleged alien creatures, the term extra*ter*restrial, meaning from outside of earth, will be replaced by extra*temp*estrial throughout this text, as the Latin root *tempus*, meaning time, is much more aligned with this cross-temporal model.

In changing the terminology used to describe the visitors themselves, it also seems fitting to put forth a new term for these "unidentified flying objects," which are likely the very devices that allow our future progeny to venture backward across the landscape of time. After all, it doesn't make sense to refer to something as "unidentified," when the purpose of this work is to identify that very thing. Additionally, those opposed to the discussion of contentious ideas, people, social movements, etc., often vilify the terms and phrases used in association with them, in an attempt to undermine their opponent's cause. "UFO," "flying saucer," "extraterrestrial," "alien," and others like them, have all fallen victim to this common social practice.

In order to overcome the indignity associated with these syntactical tactics, it sometimes becomes necessary to abandon such terms once they have become tainted by stigma. So, to help break away from the practice of subconscious dismissal of discourse as part of this cultural conditioning process, while at the same time offering up an identifiable term for a previously unidentified phenomenon, the phrase *Identified Flying Objects* will be used subsequently, and with the acronym "IFO" replacing "UFO" throughout this text. It is hoped that IFO is similar enough to the previous terminology that it may sustain a cognitive connection with the idea as a whole, while also being different enough that it may elude the shackles of shame presently associated with talking about the subject.

It is a bit curious that such a strong social taboo surrounds the subject of UFOs at all, particularly given that so many people consider it a real phenomenon. In fact, a 2013 survey conducted by YouGov and the Huffington Post asked a diverse sample of 1,000 individuals whether they "believe some people have witnessed UFOs?" Among those polled, 48% answered in the affirmative, 35% answered in the negative, and 16%

stated that they were unsure.[14] When asked about these recent results, nuclear physicist Stanton Friedman, who has worked as a Ufologist for decades, unequivocally stated that "It's always been intriguing to me how we act as though only kooks and quacks and little old ladies in tennis shoes believe in flying saucers. And it's never been true, at least for 30 or 40 years.... The believers are far more quiet, but far more on the side of reality."[15]

Another interesting result of this survey was that individuals who earned a postgraduate degree were the most skeptical group, with only 30% stating that they believe some people have witnessed UFOs. As someone with a postgraduate degree, I would appear to be in the minority here. Although I can certainly understand this result, given that the majority of us who end up going to graduate school enter a course of study that centers on rational and presently falsifiable explanations for things. However, I also believe that by assimilating established knowledge drawn from multiple academic disciplines, it may be possible to bridge this unnecessary divide between the science and science fiction of IFOs. Furthermore, it is hoped that this rigorous yet reserved academic approach can draw a more inclusive audience into the developing conversation, and initiate viable dialog among skeptics, and among those who represent various scholarly disciplines within the commonly incredulous scientific community.

1.7. Future Research into the Future Past

Throughout time, humans have conjured up faulty explanations for all kinds of odd and even entirely ordinary natural phenomena. Once our intellect and ability to test theories had developed to the point that we could begin to tackle the bigger questions, many of these false interpretations were replaced by valid scientific hypotheses. However, despite an ever-evolving understanding of the natural world and the development of technology necessary to test increasingly complex ideas, many unknowns still exist.

This book cautiously examines the premise that IFOs and extratempestrials, if real, are simply our distant human descendants, using time travel technology to visit and study us in their own evolutionary past. Presently, some elements of this idea are untestable, largely because of where we currently reside in time. However, due to significant advancements in paleoanthropology, astrobiology, astronomy, and physics over the last 50 years, we are now at least poised to begin investigating this question in real terms.

Because the human mind is so inquisitive, we must wonder if something has acted to stifle inquiry into the IFO phenomenon. Results from the aforementioned poll, as well as numerous others like it, suggest that most people have at least heard of IFOs, while a large percentage of people also accept that others may have had direct contact with extratempestrials. Anecdotally, in discussing this idea with countless individuals over the last 25 years, I have been astounded by how many have had an IFO experience of their own. This includes a fellow scientist who occasionally witnessed lights hovering above the horizon of their family ranch late at night, only to discover the following morning that some of their cows had been dissected with surgical precision.

It is rather unfortunate that this subject matter is so taboo, and that people don't speak freely about their encounters. It is also regrettable that we don't openly converse about the possibility of disc-shaped craft soaring about the heavens, or big-headed aliens picking people up to do odd things to them in the wee hours of the night. Though is it really that surprising? After all, it does sound a little crazy. Moreover, it is difficult for many outwardly intelligent people to understand even simple, well-established scientific facts—such as the process of biological evolution or why vaccines are important, for example—let alone a concept so far removed from our conventional collective consciousness.

For those who have seen an IFO or had a close encounter of any kind, the event is surely a part of their reality. Though for other members of society who lack any direct involvement, it is unquestionably more difficult to wrap their minds around such a situation, as it deviates considerably from more orthodox facets of everyday life. Indeed, for those who have had such an experience, no explanation is necessary, but, for many of those who have not, none may be possible at all.

On the other side of this same coin, it must be exhilarating for an extratempestrial granted the privilege of studying us in the past, as we represent an important erstwhile element of their own existence. As a paleoanthropologist, I can only imagine what it would be like to jet back to the past by means of a highly advanced time-traversing research vessel, pull an unsuspecting representative of *Homo neanderthalensis* out of a cave along the Dordogne River of France some 75,000 years ago, and conduct an in-depth analysis of their biology and culture. The knowledge that could be gleaned from that one short period of observational research would likely surpass all that has been acquired from the multitude of Neanderthal sites across Europe, since the first skull of this close relative of modern humans was discovered by limestone quarry workers in Kleine Feldhofer Cave, Germany, in 1856.

I have been fortunate to work at a number of archaeological sites in different parts of the world, including an *Australopithecus africanus* site in South Africa dating back to 3.5 million years ago, a Neanderthal site in southern France dating to 175,000 years ago, and numerous prehistoric Native American sites throughout Ohio and Southwest Montana. Though inevitably, toward the end of the day, while scraping back yet another 5-millimeter-thick layer of dirt to reveal a casually different piece of chert that was slightly modified from its original form, the same thought always seemed to creep back in my mind...

In the future, our anthropological colleagues may ritualistically burn their Marshalltown trowels in favor of a far more rousing and illuminating means of studying their ancestral past. Someday, our distant descendants could usher in a new era of archaeology, when we would no longer be required to sift through layers of dirt, or blast through layers of brecciated rock in order to glimpse a tiny fragment of our ancestral past. Instead, we may simply be able to spin back through layers of time, all the while garnering a far more holistic and scientific understanding of our linguistic, cultural, and biological condition, by means of the novel four-dimensional archaeological tool of time travel.

2

Why Is This Topic So Taboo and What Can We Learn by Moving Past the Stigma?

We have this propensity in our culture to ridicule UFO sightings. It's a problem, and we need to be more mature about it in this country.[16]

— Fife Symington III - Former Governor of Arizona, U.S.

It's not all aircraft lights and weather balloons and hoaxes, there is a genuine phenomenon here; something that the scientific community should want to look at. It should want to engage on this. Potentially, it has incredible things to teach us.[17]

— Nick Pope, UK Ministry of Defense, 1985 – 2006

History is filled with accounts of new phenomenon, poorly understood phenomena that have been ridiculed, and made fun of. Why? Because science hasn't had a chance to look into it or won't look into it for various reasons; we've got to turn that situation around.[18]

— Richard Haines, Ph.D. - Chief Scientist, National Aviation Reporting Center on Anomalous Phenomena

2.1. Fear and Loathing on the IFO Trail

In a poignant 1994 television advertisement for the BT Group—a British multinational telecommunications services company—physicist Stephen Hawking offered up a "message of hope," advocating for the continued advancement of innovation and technology through unrestricted communication and unfettered human dialog:

For millions of years, mankind lived just like the animals.

Then something happened which unleashed the power of our imagination.

We learned to talk. And we learned to listen.

Speech has allowed the communication of ideas, enabling human beings to work together.

To build the impossible.

Mankind's greatest achievements have come about by talking.

And it's greatest failures by NOT talking.

It doesn't have to be like this!

Our greatest hopes could become reality in the future.

With the technology at our disposal, the possibilities are unbounded.

All we need to do is make sure we keep talking.[19]

Along with the background music and dramatic imagery, the advertisement created a sense that communication was among the most important contributors to the enlightened state of our species and, potentially, to our future success as well. This message of hope was so inspiring that it was even incorporated into the Pink Floyd song Keep Talking,[20] after frontman David Gilmour was reportedly almost brought to tears by it, later stating that it was "the most powerful piece of television advertising that I've ever seen in my life, and I thought it was fascinating."[21]

I would tend to agree with Hawking that our adept language ability has been vital to the success of our species, since the early development of protolanguage beginning around 2-3 million years ago.[22][23] We could also continue to benefit from further communication regarding politics, the environment, economics, war, new research and technology, and any other matter conceivable to the human mind. As Hawking suggested in the BT Group commercial, we should advocate for open dialog about any subject, as our greatest achievements are the result of talking, while our greatest failures have come from broken discourse.

"UFOs" and "extraterrestrials" have been the subject of many science fiction books, films, documentaries, and radio shows throughout the last century. However, speaking about the subject in a non-fiction capacity tends to illicit a multitude of negative reactions. Despite this, it is still fun to imagine life on other planets and possible scenarios resulting from contact with them. For many people, the thought of life somewhere out in the deep void of space provides a sense of wonder and excitement, adding depth to our conventional conceptualization of the complexity of life on Earth.

For others, "UFOs" and "extraterrestrials" can elicit feelings of fear and uncertainty, which was perhaps most evident from Orson Welles' notorious 1938 broadcast of H.G. Wells' science fiction novel *The War of the Worlds*.[24] The unanticipated reaction to this bit of radio drama demonstrated that the perceived threat of attack by an alien life form can create a state of anxiety and mass hysteria among the easily agitated.[25] The potential for panic and chaos brought about by the perceived threat of extraterrestrial aliens is one explanation for why the topic isn't more widely discussed and why it has been repeatedly dismissed by agencies tasked with researching it. In order to govern effectively, any regime must espouse a sense of safety, control, and confidence among their people. An attack by extraterrestrials, with the potential to threaten the livelihood of those they govern, jeopardizes the authority and legitimacy of those in power.

The question of whether an extraterrestrial civilization would enslave us and steal our resources is one that has been seriously considered, even among members of the scientific community. For example, in an interview preceding a 2010 conference of the Royal Society in London, Simon Conway Morris, a professor of evolutionary paleobiology at Cambridge University, advocated that extraterrestrials who find Earth may likely be looking for somewhere to live and, if they were anything like us, they might feel inclined to help themselves to everything we have. In a shrewd jab at historical human behavior, Morris states that "Extra-terrestrials...won't be splodges of glue...they could be disturbingly like us, and that might not be a good thing—we don't have a great record."[26]

However, there are many reasons why we are not likely to face annihilation at the hands of interstellar aliens. Most notably, extraterrestrial life, which would need to be complex enough to find us in the vastness of space, does not exist anywhere close enough to Earth for it to be worth their time, energy, and effort to schlep all the way across the universe to kill us and steal our stuff. This may be the equivalent of riding a bike all the way to Pluto to smash a fly and steal a tater tot from someone.

Additionally, as pointed out in a recent article by Marcelo Gleiser, a professor of natural philosophy, physics, and astronomy at Dartmouth College, we can also take solace in knowing that if "aliens" were going to destroy human civilization, they probably would have done so already:

> From what we know now of space travel and all its difficulties—
> fuel, speed, radiation shielding—had the aliens the technologies to
> overcome all this, they would probably also have technologies that
> would have spotted us a long time ago, for example, by analyzing our
> atmospheric composition or, if closer than around 50 light years, by
> detecting our radio and TV broadcastings.[27]

In the context of the conventional model, this debate has naturally centered on the question of whether we may face the threat of annihilation by extraterrestrial aliens from a different planet. However, if the "aliens" are actually our distant progeny, it is nonsensical to think that we would, or could ever, be destroyed by those whose entire existence is dependent upon our continued survival. In fact, their extermination of us would essentially constitute a species-wide example of the "grandfather paradox," which is one of the classic time travel *consistency paradoxes* to be discussed later.

If IFOs are indeed a product of time rather than space, then the fear of attack as an explanation for the culture of dismissal that has developed around the subject is even more unfounded, given that future humans are very much stakeholders in the continued existence of their past human ancestors. So, beyond the herd panic induced by 20th century radio broadcasts and the unfounded fear of extermination by rogue aliens, what other factors may help explain the icky feeling one gets when discussing the IFO phenomenon today?

This discomfort could have come from many different places and for many different reasons. Although, reliable research, declassified documents, and history itself suggest that a primary contributor to the stigma that surrounds IFOs originated with a purposeful campaign carried out by the United States government in the 1950s. Most conspicuously, during this time, a now widely acknowledged program was put forth to quell interest in and inquiry into IFOs. Among other things, this involved propagating misinformation, epic gaslighting tactics aimed at discrediting and shaming those with legitimate experiences, and attributing these legitimate experiences to other natural or man-made occurrences.

2.2. Institutionalized IFO Inquiries

2.2.1. Manufacturing Stigma

To reiterate, I am not a Ufologist or a conspiracy theorist of any kind. I am keenly aware, however, that conversing about aliens and governmental entities in the same context acts to brand one as such and inevitably brings with it a sense of indignity and shame. In fact, part of the genius of these highly effective government-backed campaigns was to quell interest in and discussion about these very programs. Though by acknowledging what has now been divulged by these agents and agencies themselves, it is hoped that this ensconced tendency to dismiss anything associated with IFOs can be overcome, and that we can begin to mitigate the sense of stigma surrounding the subject.

Government programs such as Project Blue Book in the U.S.,[28] the Ministry of Defense and UFO Desk in the U.K.,[29] investigations carried out by the KGB in the former Soviet Union,[30] as well as the mountains of data that they and other organizations have compiled from eyewitness accounts over the years, suggest that the IFO phenomenon is worth considering seriously. These and other such organizations were largely formed in the late 1940s, during what could be considered a very active time for extratempestrials. In fact, between 1948 and 1969 the United States Air Force recorded 12,618 sightings of "strange phenomena," with 701 of these remaining unidentified to this day.[31]

The U.S. Air Force oversaw what was among the first officially sanctioned investigations of these odd occurrences. In 1948, the Project Sign (originally Project Saucer) program was established following a high-profile incident involving an IFO crash that allegedly occurred near Roswell, New Mexico, in 1947.[32] In 1949, Project Sign was succeeded by Project Grudge, which is perhaps the earliest and best place to look for the origins of the synthetic stigma that limits open dialog about IFOs today. This is because the actual stated mission of Project Grudge was to quell mounting fear among the general populous regarding perceived threats posed by unidentified objects. Additionally, Project Grudge officials were instructed to explain away these incidents and convince people that they were actually just the result of hallucinations, mass hysteria, weather balloons, or people mistaking natural phenomena for IFOs.[32]

Following a series of sightings at Fort Mammoth, New Jersey in 1951, the U.S. military revamped Project Grudge, putting U.S. Air Force Captain Edward J. Ruppelt in charge of a new agency called Project Blue Book, which was based at Wright-Patterson Air Force Base near Dayton, Ohio.[32] Captain Ruppelt was responsible for overhauling the entire project, which included developing a standardized form that could be used to file reports of IFO sightings. Ruppelt also appointed members of the Battelle Memorial Institute in Columbus, Ohio, to perform statistical analyses of these standardized reports, as part of a new program named

Project Stork. Additionally, he formed strong collaborative relationships with prominent scientists, which included Dr. J. Allen Hynek from The Ohio State University who had already been consulting for the U.S. Air Force as part of the earlier Project Sign and Project Grudge programs.[32]

2.2.2. Dr. J. Allen Hynek

Dr. J. Allen Hynek was a respected astrophysicist in the Department of Physics and Astronomy at Ohio State when Ruppelt recruited him to take on a leading role in Project Blue Book. In addition to this early role, Hynek also served as Associate Director of the Smithsonian Astrophysical Observatory at Harvard University, where he was head of its NASA-sponsored satellite tracking program. Hynek was also the director of the Lindheimer Astronomical Research Center and Chair of the Astronomy Department at Northwestern University, founded and served as the first director of the Center for UFO Studies (CUFON), authored the comprehensive book *The UFO Experience: A Scientific Inquiry,* and was even invited to speak before a General Assembly of the United Nations.[33]

As part of Project Blue Book, Hynek's efforts were put toward the development of a protocol that would help streamline reports of IFO and extratempestrial encounters. This was accomplished by standardizing variables that were important to vetting both the reports and the reporters. Among the most important contributions to this new procedure was the addition of *Strangeness and Probability Ratings*, which rated how odd or unexplainable (*Strangeness*) the encounter was and how reliable and trustworthy (*Probability*) the witnesses were, on a scale from 1 to 10. This new rating system sought to answer two distinct questions: what does the report say happened and what is the probability that it actually did happen? A benefit of this new rating system was that it allowed for easier analysis of the data. Furthermore, results of the analysis could help inform whether a specific IFO report required additional scrutiny and a more in-depth scientific investigation of the event.

As Hynek points out in his 1972 book *The UFO Experience: A Scientific Inquiry*, what determined the appropriate Strangeness Rating and how likely it was to elicit further study was not just how odd an event was, but rather, how many individual strange bits were associated with it. According to Hynek:

> A light seen in the night sky the trajectory of which cannot be
> ascribed to a balloon, aircraft, etc., would nonetheless have a low
> Strangeness Rating because there is only one strange thing about

the report to explain: its motion. A report of a weird craft that descended to within 100 feet of a car on a lonely road, caused the car's engine to die, its radio to stop, and its lights to go out, left marks on the nearby ground, and appeared to be under intelligent control receives a high Strangeness Rating because it contains a number of separate very strange items, each of which outrages common sense.[33]

At that time, as it is now, Hynek understood that most people considered IFO reports to be the sensationalized stories of half-drunk backwoods hicks. After all, part of his role while working for Project Sign and Project Grudge was to explain away sightings as the result of natural phenomena, while also discrediting the observer. This proved to be a highly effective method of discrediting those who had a close encounter, given that one of the best ways to delegitimize legitimate factual reports is to make the people conveying them seem insane or untrustworthy. In this role, Hynek was an instrumental component of the debunking campaign, acting as the rational scientist who was put forth to offer up a worldly explanation for a seemingly other-worldly occurrence. This was a task that was expected of him, but it was also one that he rather enjoyed during these more skeptical early years, at least until he began to accept this phenomenon as real in later life.

As Hynek continued to collect greater amounts of data, he started to realize that the majority of people who filed these reports were just average people who happened to have had a very strange encounter, which for many, had been a life-altering event. With this in mind—and with the added benefit of hindsight—Hynek's role as part of Projects Sign, Grudge, and Blue Book was somewhat paradoxical. He helped create this easily dismissible stereotype of a crazy, toothless, drunk person who gets abducted from a remote wooded shanty and anally probed. However, he was also tasked with trying to accurately assess whether these individuals were in fact the crazy people his programs were making them out to be, with the hope of gaining a better understand of the true nature of their experiences.

The Probability Rating was instrumental in helping to differentiate between the fake or embroidered accounts and those given by credible people who witnessed something that couldn't be explained with a contemporary understanding of technology, space, and time. Although to Hynek, assessing one's personal disposition was seen as a more subjective part of the scientific inquiry, given that it was essentially a short-term character assessment of a complete stranger. However, in addition to the personal credibility assessment, other factors were also considered, including the internal consistency of the report, consistency across several separate reports of the same incident, how the report was made, the conviction of the individual as they recounted events, as well as a general assessment of how it all fit together.[33]

Hynek also realized that it was important to differentiate among the specific types of experiences that people had, which led him to develop another classification scheme involving three hierarchical categories of what he referred to as *Close Encounters*. According to Hynek, a Close Encounter of the First Kind involves an IFO observed at some distance, or in the form of "Daylight Discs," "Nocturnal Lights," or "Radar/Visual Reports." Close Encounters of the Second Kind were those observed at a closer range, generally within 500 feet, which allowed for more details to be observed and described. Lastly, Close Encounters of the Third Kind referred to instances in which actual animated creatures were observed in or around the IFO.[33]

Hynek ended the classification hierarchy at simple observation of the craft's occupants, and did not include reports involving individuals who claim to have been abducted by them. However, in a 1998 paper published in the *Journal of Scientific Exploration*, Jacques Fabrice Vallée, one of Hynek's close associates, argued that Close Encounters of the Fourth Kind should also be added to the existing scheme. This was meant to account for individuals who reported being abducted, as well as those who experienced an alteration of their sense of reality in association with an IFO encounter.[34] Later, Steven M. Greer, founder of the Center for the Study of Extraterrestrial Intelligence (CSETI), recommended that Close Encounters of the Fifth Kind also be added to account for instances of intended, conscious, compliant communication with an "extraterrestrial intelligence."[35]

Hynek's work was important in that it established a foundation upon which a better understanding of IFO and extratempestrial encounters could be built. The development of a standardized quantitative system was critical for assessing the validity and credibility of IFO reports that have and continue to occur throughout the world. His work also helped pave the way for subsequent large-scale investigations of this phenomenon, including a panel of esteemed scientists (under the direction of physicist Dr. H. P. Robertson of the California Institute of Technology) who were asked by the Central Intelligence Agency to analyze data compiled as part of Project Blue Book.

This group of scientists, which became known as the Robertson Panel, made recommendations regarding the validity of IFO reports and identified potential threats that they may pose to national security.[32] The Robertson Panel spent two weeks examining these data and issued their findings as part of the *Report of Scientific Advisory Panel on Unidentified Flying Objects Convened by Office of Scientific Intelligence, CIA* January 14-18, 1953, or in short, *The Durant report of the Robertson Panel proceedings*.[36] In it, they concluded that IFOs were not a direct threat to U.S. national security. However, and in spite of any actual threat, the panel advocated that an effort should be made to diminish the perceived prominence of IFOs in order to subdue public interest in the matter.

As a result of its considerations, the Panel concludes:

> 3.a. That the evidence presented on Unidentified Flying Objects shows no indication that these phenomena constitute a direct physical threat to national security.

> 4.a. That the national security agencies take immediate steps to strip the Unidentified Flying Objects of the special status they have been given and the aura of mystery they have unfortunately acquired. ... We suggest that these aims may be achieved by an integrated program designed to reassure the public of the total lack of evidence of inimical forces behind the phenomenon, to train personnel to recognize and reject false indications quickly and effectively, and to strengthen regular channels for the evaluation of and prompt reaction to true indications of hostile measures.[36]

This recommendation by the Roberson Panel was instrumental in shaping what would become a much more strident effort aimed at separating this phenomenon from the reality of mainstream American culture. As shown above, an explicit debunking campaign had already been put in place as part of Project Sign and Project Grudge. However, the decrees put forth by the Robertson Panel initiated a much more draconian strategic plan aimed at manufacturing stigma surrounding the subject of IFOs, which has persisted with fervid tenacity in American society since this time.

2.3. International Institutionalized IFO Inquiries

In the 1950s, the Canadian government began conducting its own investigation of IFOs as part of what was known as Project Magnet. Much like in the U.S., this research was the result of numerous sightings and reports of "flying saucers," observed throughout the country at that time. Project

Magnet was spearheaded by Wilbert B. Smith, a senior radio engineer from the Department of Transport. In 1950, Smith asked his superiors for permission to use the department's laboratory facilities and equipment to study the physical properties of IFOs, with an emphasis on examining the propulsion system of these craft.[37] According to records kept by the Canadian government's Library and Archives Canada:

> The goals of Project Magnet were fueled by the concepts of geomagnetism, and the belief that it may be possible to use and manipulate the Earth's magnetic field as a propulsion method for vehicles . . . Smith believed he was on the "track of something that may prove to be the introduction to a new technology.[38]

From descriptions of Smith's work, which took place around the same time as Project Grudge and Project Blue Book in the U.S., it is clear that the Canadians were also very interested in IFOs. However, relative to the U.S. government, Canadians had a fundamentally different way of going about investigating them. For instance, U.S. government agencies were actively trying to quell interest in the subject, and they worked tirelessly to debunk reports provided by those who had seen an IFO. However, in Canada, people who had witnessed these advanced flying machines were seen as an asset, and researchers were interested in what they had to say about them.

According to documents at the Library and Archives Canada "It was believed by both Smith and other government departments involved, that there was much to learn from UFOs. Investigations into these sightings and interviews with the observers were the starting point for Project Magnet."[37]

In addition to Canada's more inclusive approach to investigating the IFO phenomenon, their reports were also presented with far more conviction than those in the U.S. In fact, Smith's 1952 Project Magnet report overtly acknowledged the reality of what people were observing in the skies above Canada. Through a comprehensive analysis of vetted reports, he was even able to describe specific characteristics of these craft and concluded that they were not associated with any known Canadian technology, or likely any technology that existed anywhere in the world at that time:

> In a summary of 1952 sighting reports, Smith noted common significant characteristics of UFOs: "They are a hundred feet or more in diameter; they can travel at speeds of several thousand miles per

hour; they can reach altitudes well above those which should support conventional air craft or balloons; and ample power and force seem to be available for all required maneuvers....Taking these factors into account, it is difficult to reconcile this performance with the capabilities of our technology, and unless the technology of some terrestrial nation is much more advanced than is generally known, we are forced to the conclusion that the vehicles are probably extra-terrestrial, in spite of our prejudices to the contrary."[39]

Outside of North America, numerous other sovereign nations also began setting up government agencies to conduct their own investigation of IFOs. In each case, these became necessary to deal with the flood of reports that were coming in from all over the country. For example, The Belgian Society for the Study of Space Phenomena was established in 1971 and was set up to document a mass of reports provided by both civilians and high-ranking military personnel. This was particularly true in 1989 during what was dubbed the "Belgian Flap" or the "Belgian Wave," which involved hundreds of eyewitness reports, videos, and photographs of strange, flat, silent, low-flying craft in the skies above Belgium.[40]

The existence and intent of these government-sanctioned agencies is no longer a secret and much has been divulged about the early stages of their IFO research. It is also clear from recently declassified documents produced by these previously classified agencies that the manifest function of many of these programs, at least outside of Canada, was to guide public perception of the phenomenon. Unfortunately, this practice acted to lay an early foundation upon which the ensuing culture of dismissal, scorn, and ridicule was built. With a current awareness of this manipulative intent to suppress and mislead, perhaps it is time we vanquish the synthetic stigma that has long-prohibited open dialog about this otherwise interesting phenomenon.

Additionally, while recognizing these overt institutionalized efforts to stifle inquiry and undermine the credibility of those who have had a credible experience, it is also important to consider the ubiquity, longevity, and intent of these government agencies. For instance, the sheer number of these enduring programs throughout the developed world—established for the sole purpose of investigating IFOs and extratempestrials—suggests that something was/is indeed going on. In other words, it is not likely that these organizations would be developed, persist for so long, and work so tirelessly to quell interest in the topic, if there was not actually something happening that required secretive state-sponsored inquiry. After all, there are no government-sanctioned Sasquatch Investigation Units, Unicorn Patrols, or Poltergeist Research Panels.

2.4. How Extratempestrials Differ from Sasquatch, Apparitions, and Other Fabled Phenomena

Copious reports of highly advanced craft, with technology far beyond that of modern human capabilities, have contributed to the development of institutionalized agencies tasked with investigating them. The ubiquity of these government agencies would seem to suggest the reality of IFOs and extratempestrials. Though one could also argue that some of these encounters were just the result of a faulty collective consciousness that had been built up around the phenomenon. This psychosocial tendency occurs when people who are aware of something look to impose that awareness on things that can otherwise be easily explained as something else.

With regard to IFOs and extratempestrials, this is often referred to as the Psychocultural Hypothesis. More specifically, it attributes IFO reports to a general awareness of the commonly described "alien" form and states that people who are aware of this form falsely impose it on things that are not actually aliens at all. This psychosocial proclivity can also involve the misappropriation of a real phenomenon, in which a tangible event/occurrence is incorrectly applied to one that is less perceptible. For instance, the beginning of human spaceflight may have caused us to imagine that there must be other advanced civilizations in the universe that were also entering a phase of spaceflight, which would lead us to begin looking for evidence of these expat adventurers on Earth, whether they exist or not.[41][42]

This cultural-consciousness interpretation often reminds me of a story that one of my graduate thesis advisors once told, in which he was repeatedly contacted about a Sasquatch in the Appalachian hills of Tennessee where he was living at the time. However, once the "Sasquatch" in question was captured, it turned out to be a baboon that had escaped from a nearby primate sanctuary. Because people do not expect to see an African primate frolicking around in the hills of Tennessee, they were quick to brand the thing they were confused by as a mythical creature, despite there being a more credible explanation for it.

There is no doubt that some number of IFO reports are the result of this psychocultural propensity. However, given the credentials, skills, and experience of countless others who have also experienced a close encounter, there is cause to believe many of these reports. For people like myself who have never seen a saucer-shaped craft, it is easy to be skeptical. However, because so many credible witnesses have observed an IFO or its occupants at close range—including pilots, police, scientists, and government and military personnel—it becomes harder to dismiss this phenomenon as merely the manifestation of a feeble collective consciousness.

It is easy to lump IFOs and extratempestrials in with the Loch Ness monster, Sasquatch, vampires, ghosts, and even extraterrestrial aliens. However, while it is unlikely that we will ever find undeniable evidence of these other entities, if the IFO phenomenon is in fact our time traveling descendants, then the likelihood of irrefutable contact with them is not only high, it is also inevitable, as this would mean that we are destined to someday become them. It is for this basic reason that the question of time traveling humans is profoundly different from that of whether extraterrestrial aliens may be visiting us from a different planet, or whether the Loch Ness monster, unicorns, or yeti exist for that matter. In other words, the current time-travel hypothesis makes a specific prediction about a specific future human condition, which can be tested simply by continuing to exist into that future. In fact, even if we were to annihilate ourselves at some point in the coming years, the hypothesis will still have been tested... and rejected, despite there being no one left to know.

I would reluctantly admit that there is a minute chance that a Sasquatch-type creature that is not a baboon could potentially exist in a heavily wooded and very remote region of Earth. I highly doubt that this fabled animal is actually real, but if it were, then we would expect to find it eventually. However, with each jump in the global human population and with each tree that we cut down in the forest, the chance of discovering a large hairy Sasquatch somewhere in the finite expanse of this planet grows ever slimmer.

Furthermore, after some period of time, when humans have expanded to all parts of the Earth, if we still haven't found a living Sasquatch anywhere in the world then we must inevitably conclude that they don't actually exist. This same caveat also holds true regarding the question of extraterrestrials from another planet. We could wait for an eternity, while continuously scouring the heavens above, looking for some sign of alien life in the great beyond, and never find anything at all. If this were to occur, at some point we would have to surrender to the fact that we may be the only intelligent lifeform in our celestial neighborhood.

Despite the low probability that we will ever find intelligent extraterrestrial humanoid life, we continue to spend huge sums of money looking for it. In fact, on July 20, 2015, the recently departed physicist Stephen Hawking—whose compelling quote began this chapter—hosted a news conference in London, England to announce that he and Russian tycoon Yuri Milner would be spearheading a $100 million project called "Breakthrough Listen," dedicated to searching for intelligent extraterrestrial life in our galaxy.[43]

Scanning the skies for alien radio broadcasts and sending biographical data about our own species out into space—though trifling given the magnitude of that which lies beyond Earth—is the best we can currently do in an attempt to find whatever it is we think we are looking for. If the goal is simply to know what, if any, intelligent civilizations may lie beyond the boundaries of our own solar system, then this is certainly a noble pursuit. However, if we are seeking answers to the question of IFO sightings and accounts of alien contact, then I would argue that we are not only looking for the wrong thing in the wrong place, but also in the wrong time.

Admittedly, we still lack concrete, publicly available, undeniable evidence of the existence of extraterrestrials or extratempestrials. There are no pictures or videos of time traveling humans hanging out with any leader of the modern world, for example. However, in spite of this lack of hard evidence, we may potentially find indirect evidence of cross-temporal contact in human prehistory, history, and at present, in the form of consistent data compiled by respected individuals and agencies who have actively researched this phenomenon for over 70 years. Furthermore, these recent accounts are frequently provided by well-educated, well-respected, sound-minded individuals, who often occupy academic, government, and military positions. Though in order to reap any potential benefits that could be gleaned from investigating instances of intertemporal interaction, we must first move past the indignity associated with discussing the topic, and instead embrace a more open-minded, holistic, and collaborative approach to examining the IFO phenomenon...all we need to do is make sure we keep talking.[44]

3

A Brief History of Time . . . Travelers

Skeptical scrutiny is the means, in both science and religion, by which deep insights can be winnowed from deep nonsense.[45]

– Carl Sagan

3.1. The Question of Extratempestrial Encounters in Human Prehistory

The dominant worldview regarding IFOs has long been that alien creatures from a different planet are visiting us here on Earth. More recently, numerous speculative ventures have emerged that attempt to interject extraterrestrial influence into a plethora of cultural remains associated with the historic and prehistoric human past. Various claims center on the notion that interplanetary visitors helped build some prominent megastructures and megalithic sites around the world, such as the Egyptian pyramids, Machu Picchu, Carahunge, Easter Island, and Stonehenge, for example. The frequently espoused view among self-described "ancient alien theorists" is that these massive structures, as well as numerous other large features, could not have been built by these ancient peoples. Instead, they advocate that these features must have been erected by a much more advanced extraterrestrial civilization that had visited Earth during those times.

However, the field of Experimental Archaeology—which uses experimental methods to test hypotheses about how artifacts and features were produced—has consistently demonstrated that these types of megalithic sites could certainly have been constructed with a large and well-organized human workforce.[46][47][48] Additionally, numerous other examples of major public works projects, accomplished with only slightly more advanced technology (and often a fair amount of slave labor), can be found in these same regions not long after the megalithic and monumental features were erected, which further elucidates the earthly abilities of these past human groups.

Beyond these megalithic structures, numerous other examples of art, artifacts, body modification, etc., are commonly explained as being made by or the result of extraterrestrial influence. For example, the *Saqqara Bird* from Saqqara, Egypt, which is dated to the second century B.C., somewhat resembles, to a very minor extent, a modern-day airplane or glider (figure 3-1). Because of this, some fringe interpretations have emerged asserting that it reflects past contact with advanced flying machines.[49]

Figure 3-1. *Image of the 2,200-year-old Saqqara Bird from Saqqara, Egypt.*

Although it is impossible to know the inspiration behind objects such as these—merely because human intention is not preserved in the archaeological record—it is far more likely that this, and other commonly cited examples of "flying machines," are simply attempts at making something much more conventional. This is particularly true of the *Saqqara Bird*, which really just looks like a bird that someone tried to carve out of wood a very long time ago.

Any misinterpreted "plane" or "glider-like" elements of this carving likely only reflect the difficulty of making a bird out of wood—and particularly a bird's wings—using crude instruments over 2,000 years ago. Additionally, IFOs don't look anything like airplanes, so it doesn't make sense why someone would make this carving look so much like a bird, if they were actually trying to recreate the form of a saucer-shaped craft.

3.1.1. Artifacts and Features, Space and Time

In spite of certain members of the ancient aliens camp occasionally overstating the physical and symbolic qualities of past artifacts, some cultural remains would seem to possess certain characteristics that may warrant further scrutiny. Additionally, the ubiquity of artifacts, features, and artistic representations of scenes depicting more disc-shaped craft and humanoid forms with large rounded heads for example—which exist broadly across different societies and over multiple time periods—raises more legitimate questions about the possibility of past contact.

However, if certain relics of the past reflect the influence of an advanced civilization, then the more parsimonious explanation (i.e., that which uses fewer assumptions and conjectures) is one of intertemporal interaction between our future human descendants and past human ancestors, as opposed to contact with an extraterrestrial civilization. After all, the only evidence of an increasingly advanced civilization anywhere in the universe exists in the form of 3.3 million years of progressively sophisticated material culture as it evolved across sequential human groups, right here on planet Earth.

Figure 3-2. *Image of the "Owl Man" or "Astronaut" geoglyph at Nazca, Peru, as viewed from high above.*

Considering possible indications of cross-temporal contact in the pre-historic archaeological record, one interesting case may be found among a group of geoglyphs that were built by the Nazca people of South-west Peru, between 1,200 and 2,100 years ago. Geoglyphs are large-scale designs or motifs that are created by adding or removing objects in an environment to distinguish between the design and the naturally occurring landscape. The Nazca people erected a number of different geoglyphs during the height of their civilization, though one dubbed the "Owl Man" or "Astronaut Man" tends to stand out among them (figure 3-2).

Many interpretations have been put forth to explain this large feature. However, much like the *Saqqara Bird* and countless other prehistoric artifacts and features, no one can know the true reason why it was made, as the culture and history of the people that created it have not fully endured into modern times. With this limitation in mind, some aspects of these geoglyphs—and particularly those associated with the "Astronaut Man"—do raise questions about possible contact between future and past human groups.

However, it is important to reiterate that this geoglyph does not constitute any form of proof regarding the proposed time-travel hypothesis, and again, we must remain vigilant in separating information from interpretation and science from suggestion. Though in questioning why a group of people would go to such great lengths to create a 100-foot-tall image of a big-headed, big-eyed, human-like being, etched across the face of a steep mountain on the Nazca plateau, it is perhaps worth considering as a possibility. This may also be a worthwhile pursuit considering that contact with, wonderment of, and a likely perception of extratempestrials as gods, would be expected to manifest itself in this and similar other forms throughout the past.

3.1.2. Intentional Cranial Modification in Human Prehistory

Beyond art, artifacts, and features associated with human prehistory, the process of cranial modification—the intentional binding of an infant's skull—is another ubiquitous aspect of the past that raises questions about possible intertemporal contact. It should be noted however, and contrary to the beliefs of some ufologists, that past persons who had their heads bound as an infant were not actual aliens, nor were they the result of interbreeding between humans and an alien race, whether it be from space or time. Rather, this was an extremely common type of body modification, which existed throughout the world and across long stretches of the historic and prehistoric past.

Cranial modification was carried out for many different reasons, and because of widespread variation in why this practice occurred it should not be considered broad-based evidence of contact with extratempestrials.

However, ethnographic research has revealed that certain cultural groups engaged in cranial modification because they were instructed by, or they wished to mimic god-like beings. Additionally, most forms of cranial modification result in a taller and broader forehead, more prominent and protruding eyes, and a smaller and more retracted face. As such, in some cultural groups, this practice could have been influenced by a desire to look like our time traveling descendants, who are likely to possess such characteristics as a result of their further evolutionary development through time.

Figure 3-3. *Example of a common form of head binding used by the Maya.*

Intentional cranial modification is a process that typically involves placing a tight wrap around the neurocranium—the part of the skull encompassing the brain—at a time shortly after birth, when an infant's developing calvarium (skullcap) is still highly malleable (figure 3-3). Because of integration among bones that comprise the craniofacial skeleton and because the brain and braincase are the first anatomical regions of the skull to grow in humans, binding the cranium elicits an effect on downstream anatomical regions that develop after those of the upper skull.[50 51 52 53 54 55] For instance, compressing an infant's neurocranium acts to create a higher and wider forehead and cranial base, while also reducing the amount of time and space available for facial growth, which results in a more orthognathic viscerocranium, meaning a more retracted facial skeleton (figure 3-4).[51 56 57]

Applying pressure to the posterior and inferior neurocranium in the region of the occipital lobe at the back of the skull acts to shorten cranial length and limit anteroposterior space for the eyes and orbits that surround them. In other words, binding the skull causes it to shorten in the sagittal plane (i.e., front to back), which reduces the depth of the eye sockets and other structures in this anatomical plane. In association with this form of cranial modification, the eyes become more exophthalmic (meaning that they protrude forward beyond the margins of the orbit), which causes them to appear larger and stand out more prominently among other cranial and facial characteristics.

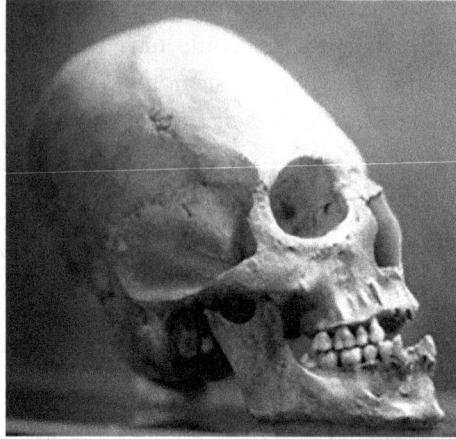

Figure 3-4. *Example of the result of this process observable in the craniofacial anatomy of an adult human dry skull.*

As stated above, the collective result of most forms of intentional cranial modification is an accentuation of prominent craniofacial trends that have occurred throughout the last six million years of hominin evolution, and which are likely to persist throughout the human future as well. It is currently impossible to know exactly what we will look like in the distant future. Though if the same enduring trends of facial reduction, brain expansion, and neurocranial globularity continue to persist, our distant progeny may possess a morphological form similar to those who have been subjected to cranial deformation, which itself may have been carried out in an effort to mimic these future human morphological characteristics among certain cultural groups.

Intentional cranial modification is an ancient practice. In fact, the earliest indication of cranial deformation is associated with two Neanderthal skulls recovered from Shanidar Cave in modern day Iraq, which date back to at least 45,000 years ago.[58] The procedure was also widespread within and among prehistoric and historic groups, having been documented in almost every geographic region of the world (figure 3-5),[59][60][61] including an estimated 89% of the Mayan population of Mesoamerica.[62]

Many explanations have been put forth regarding why this practice originated in so many different societies throughout so much of the human past. Among the more academically rigorous interpretations of this

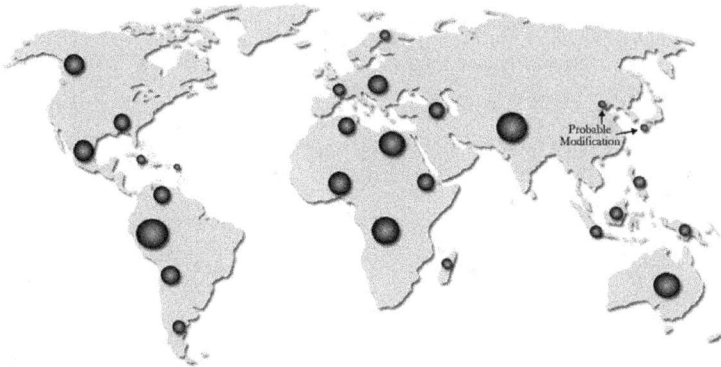

Figure 3-5. *Major geographic regions where cranial deformation was practiced.*

* Image generated using data compiled from existing literature on cultural and archaeological indications of intentional cranial deformation among different regional societies throughout the world.

behavior are that it was a form of tribal identity, a cultural or ethnic boundary marker, imitation of certain animal forms, designation of nobility status, imitation of nobility status, to appear fiercer during warfare, and as mentioned above, because their ancestors were instructed by the gods to perform these cranial deformations. While the latter explanation may stand out as a fringe-interpretation of the practice, this gods-based justification has actually been revealed by and echoed throughout a number of different ethnographic studies. For instance, according to Gerszten and Gerszten (1995), "Many living people who have undergone intentional head deformation claim that their gods instructed their ancestors to perform the practice and that they are simply fulfilling the desires of their gods."[60]

While a more conventional interpretation of this practice centers on tribal and class distinctions, it is interesting that some groups claim the gods told their ancestors to do it, or that they were attempting to mimic the form of their gods. Again, this in no way proves that our time-traveling descendants intentionally or unintentionally influenced the practice of cranial modification. Though considering just how widespread this independently invented custom was, the pain and discomfort inflicted upon infants to illicit these craniofacial effects, as well as the end result of cranial binding itself, it is perhaps worthwhile to at least consider this custom in the context of potential intertemporal interaction.

Because intentional cranial modification results in clear morphological changes, this could represent a more tangible manifestation of the experiences and belief systems of these cultures. Again, we cannot currently know what true meaning was associated with this procedure. Although, if cross-temporal contact did occur in the places and times that this practice originated, it could perhaps be interpreted as an attempt to recreate the craniofacial morphology of these more biologically and technologically advanced humans of the distant future.

This form of craniofacial mimicry would also be expected if past peoples perceived future peoples as gods, due in part to their more advanced technology, language, style of dress, and distinctive morphological features. Moreover, even without ongoing contact through time, this practice, as well as its stated association with the gods, would become an integral part of the myth, folklore, and religion of the descendants of these earlier cultural groups. As such, cranial deformation would persist, as it often has, simply as a result of cultural inertia among subsequent generations, long after the initial purpose and meaning behind it was lost.

3.2. Cynicism of Eccentric Antiquity

Caution is warranted with any interpretation of the symbolic meaning behind any aspect of prehistoric material culture, including intentional cranial modification and other forms of body alteration. Realistically, artifacts, features, and customs of the ancient world could have served any number of religious, aesthetic, sexual, economic, or copious other individual or societal functions. A high level of variability in the meaning behind material culture is also revealed through ethnographic and ethnoarchaeological research, in which the living descendants of past peoples are able to provide insight into the various symbolic, stylistic, and functional elements of the material culture left behind by their ancestors. Without this insight, these vestiges of the past may also be subject to erroneous interpretation.

As an example, on the island of Yap, in the Caroline Islands of the western Pacific Ocean, giant stone rings known as Rai stones dot the landscape. These megalithic features, occasionally weighing upwards of 8,000 pounds, were quarried over 1,500 years ago. Because of their antiquity, massive size, and unique shape, these stone rings could be interpreted in a multitude of different ways. However, ethnographic research has revealed that these actually served a vital economic function, acting as an important form of non-portable currency, or money on the island. Without the ability to ask the indigenous inhabitants of Yap about the symbolic and functional meaning behind these Rai stones, we would be left to speculate about their purpose, as we often do with other ancient relics of erstwhile cultures. Additionally, it isn't likely that we would ever correctly guess their true function as a form of currency, given how divergent it is from our own modern form of money.

Stonehenge, Carahunge, Easter Island, and other megalithic structures of the ancient world were certainly erected by the indigenous people that inhabited these regions and times. Though, why they were built and what specific meaning they had cannot be known without the ability to ask directly. As such, these types of artifacts and features tend to lend themselves to a wide array of untestable interpretations. However, it should be noted that members of the "alternative community" are not the only ones guilty of conjecture, as even accredited academics occasionally put forth untestable theories to explain past human behavior.[63] In fact, given the nature of anthropological research, perilous interpretations of archaeological and paleoanthropological data are pervasive. Additionally, opposing opinions frequently result in highly contentions debates at scholarly conferences, in the scientific literature, and more recently, in various mass media outlets.[64]

Regardless of the source, it is important to assess the merit of any interpretation of past human morphology, behavior, or material culture. We must also recognize that it is all too easy to add something that is completely out of the ordinary when there are gaps in our conventional knowledge, and particularly when it is impossible to verify or refute the claim. In his quote that began this chapter, Carl Sagan points out that skeptical scrutiny can help tease apart deep and insightful thoughts from those rooted in deep nonsense. Sagan was certainly a free-thinker but advocated for a cautious and critical form of open-mindedness. In fact, he was fond of saying that "It pays to keep an open mind, but not so open your brains fall out."[65] He also acknowledged how important it is to the advancement of knowledge and science that we strike a balance between overly free and overly skeptical objective thought.

> Science is much more than a body of knowledge. It is a way of thinking. This is central to its success. Science invites us to let the facts in, even when they don't conform to our preconceptions. It counsels us to carry alternative hypotheses in our heads and see which best match the facts. It urges on us a fine balance between no-holds-barred openness to new ideas, however heretical, and the most rigorous skeptical scrutiny of everything—new ideas and established wisdom. We need wide appreciation of this kind of thinking.[66]

It is vital that we remain open-minded, while critically assessing that which enters the open mind, judiciously adjudicating the source of information and any potential biases that may be associated with it. The IFO phenomenon should also be viewed pragmatically, and we must practice skeptical scrutiny in assessing various claims about the past and present. However, we must also acknowledge when something does not clearly conform to the known skillset of people from a specific time period, or when a cultural group exhibits traits that may reflect outside influence. Taken together, more answers may come from a skeptical open-minded approach to the possibility of future bridges being built to the past, which may help shed some light on the more peculiar physical and sociocultural elements of the prehistoric, historic, and recent human past.

3.3. The Question of Extratempestrial Encounters in Human History

3.3.1. Anomalous Events in the Historical Record

Without a time machine, the intent, symbolism, and meaning behind prehistoric art, artifacts, and features cannot be known, as their cognitive origins are not preserved through time. This puts limits on our ability to fully understand the deep past and leaves the door open for speculative and untestable interpretations of the customs and cultural remains of early societies. However, following the advent of writing and with more material available for analysis from recent history, the beliefs, customs, and behaviors of historical societies can provide a clearer picture of the past.

Not all written and verbal accounts of historical events are factual, as many are fictitious narratives rooted in origin myth, legend, and folklore, while some were produced to "god the gaps" in knowledge, and others were created merely for the purpose of entertainment, as is common today. However, oral legends and written accounts produced by indigenous societies that were based on real events can be valuable, in that they provide a rare glimpse into a cultural past that has become eclipsed by time.

Among these oral and written accounts can be found a multitude of stories describing flying machines, bright lights, and odd-looking anthropomorphic creatures descending from the skies above. Based on our own conceptualization of reality, these are easy to write off as the result of an overactive imagination or a hallucinogenic shamanistic vision. However, it is possible that many of these accounts are real and that they stem from fleeting bridges being built to connect disparate human cultures, formerly separated by vast swathes of time.

Past humans who happen to encounter future humans would undoubtedly find the culture, technology, and general appearance of the latter odd, simply because the characteristics of these more advanced humans from the future exist far beyond that which has ever been known in earlier times, or now for that matter. Even though the biocultural characteristics of these future humans are inevitably built upon ancestral elements of the past humans, they are likely to be unrecognizable to any past human group, including our own, because of diversity among cultures and the speed at which culture changes over time. Because of this chrono-cultural disconnect, such encounters—and the way in which they were chronicled in the historical record—may seem fictitious and unbelievable, even if they were the result of actual events.

Misinterpretation of the physical and cultural traits of groups from different time periods is certainly a unidirectional affair, given that individuals looking back from the future already have some knowledge of the past. For instance, if someone from the year 1369 were to suddenly show up driving a horse carriage and wearing 14th-century clothing, it would not challenge our existing notion of and familiarity with the perceived flow of time. Moreover, even if that person were truly from that time—because we are familiar with this stage of human history—it would be far less shocking than someone descending from the sky in a spinning disc and performing medical examinations on us using advanced and unrecognizable tools of the trade.

Numerous oral and written accounts of extraordinary events could have originated from close encounters with IFOs, particularly considering the large amount of detail provided in many of them, which isn't likely to have been spontaneously generated without some catalyzing occurrence. One example of an anomalous event in the historic past, which was witnessed and described by a large group of people—thus earning it a higher Probability Rating on the Hynek scale—occurred in 1803 near a small coastal village on the island of Japan. This incident is known as *The Legend of the Hollow Boat (Utsuro-Bune)*, and a number of complimentary accounts of it have been archived in a text called the *Hyouryuukishuu (Tales of Castaways)*, which is housed at the Iwase Bunko Library in Japan.[67] The legend of the Utsuro-Bune describes an encounter in which a strange saucer-shaped craft washed ashore near the small village of Harashagahama on the coast of Japan (figure 3-6).

Figure 3-6. *Image of the craft and its occupant, as drawn by witnesses following the 1803 Utsuro-Bune event.*

It was reported to be approximately five meters wide and three meters tall, was made of shiny metal, and it had a peculiar form of writing in an unrecognizable language inside the vessel. Witnesses also recounted seeing a beautiful young woman inside the craft who spoke an unknown language and carried a small box that no one was allowed to touch.[68]

If this was an actual historical event, the characteristics of both the craft and human suggest that this may have been an unintentional encounter between people from our recent past and distant future. For instance, the female inside this elliptical vessel was clearly not from Japan at that time, considering that she was piloting something that was not a wooden boat or ox cart and because she reportedly had a pale face and reddish hair, which are not typical of that region and time period. Additionally, as will be discussed in subsequent chapters, the disc-shaped characteristic of this and other IFOs may represent one of only a few shapes that facilitate time travel to the past.

Among the many odd elements of this encounter, what is perhaps most interesting is that the woman's status as a human being was never brought into question, despite some differences between her and the indigenous population of Japan at that time. This indicates that she was similar enough to these people that she remained recognizable as a member of their same species, but that she and the vessel she arrived in were different enough that they warranted written documentation of the event. Had she instead arrived in a canoe and emerged as an 8-legged man-eating spider monster, the legend of the Utsuro-Bune, as well as this time travel interpretation of it, might be slightly different.

In addition to the legend of the Utsuro-Bune, the vast majority of reports of close encounters also describe extratempestrials in very human terms, though with some variation our heads, eyes, skin tone, and height. However, given the frequency with which modern and historic accounts describe bipedal humanoid creatures piloting advanced elliptical airborne craft, it is somewhat surprising that the dominant mantra remains one of space travel over time travel, and that we don't ask "why is that human in a flying saucer" as opposed to "what planet did that alien come from?"

Legends, folklore, origin myths, and other historical accounts of strange phenomenon certainly do not all stem from actual events. Although, given the frequency with which ethereal experiences are described in oral and written narratives of the past, there may be value in keeping an open mind to the possibility that certain sagas originated with our recent human ancestors observing more advanced human groups from our shared future. Peculiar accounts of IFOs from the recent and historic past compel further scrutiny, as they, along with current indications of intertemporal interaction, may offer deeper insight into the past, present, and future human condition over much larger tracts of time.

3.4. Recent Indications of Anachronous Encounters

Copious accounts of IFOs have been documented in modern times, and numerous indications of contact exist throughout the past. However, definitive proof of extratempestrials remains elusive. This would seem to suggest that intertemporal inquiry is a somewhat one-sided affair, with knowledge transferred in a past-to-future direction, but not the other way around. Of course, it could also be argued that this lack of concrete proof indicates that this is not a real phenomenon.

Those who advocate for the existence of interstellar space aliens often use this lack of hard evidence as a chance to invoke the commonly stated aphorism "the absence of evidence is not evidence of absence." This is referred to as the *Argumentum Ad ignorantiam*, meaning the *Appeal to Ignorance*, in which ignorance represents a lack of evidence to the contrary. The appeal to ignorance is a type of fallacy of informal logic, which, in its most basic form, advocates that because of a lack of evidence, a claim must be either true or false, depending on the argument being made. In his book *The Demon-Haunted World*, Carl Sagan lists the appeal to ignorance as an important component of what he calls the "baloney detection kit," as it is one of 20 such fallacies that everyone should be cautious of in approaching questions regarding the unknown.

> *Appeal to ignorance*: the claim that whatever has not been proved false must be true, and vice versa. (e.g., There is no compelling evidence that UFOs are not visiting the Earth; therefore, UFOs exist, and there is intelligent life elsewhere in the Universe. Or: There may be seventy kazillion other worlds, but not one is known to have the moral advancement of the Earth, so we're still central to the Universe.) This impatience with ambiguity can be criticized in the phrase: absence of evidence is not evidence of absence.[69]

In this book, Sagan discusses the appeal to ignorance fallacy with regard to the existence or nonexistence of extraterrestrials, which differs from the question of time travel and human evolution. However, a cautious approach to this logical fallacy is also warranted in the context of IFOs and extratempestrials. For instance, one cannot simply conclude that extratempestrials do not exist without hard evidence of them in our time, nor should we conclude that they do exist, simply because there is no evidence to suggest that they do not. Rather, an abductive approach and skeptical assessment that considers all available evidence from the human past, present, and future may be best, given our current inability to travel through time in order to see for ourselves.

Taking circumstantial evidence into account also helps mitigate some problems associated with the appeal to ignorance, given that ignorance cannot exist when circumscribed by abundant evidence. However, designating circumstantial evidence as evidence in any form is also up for debate, due to our more limited vantage point resulting from where we currently sit in time. With this limitation in mind, less concrete forms of evidence, such as sightings, images, videos, etc., would be expected to shift from merely circumstantial to more tangible forms of actual evidence, as we draw nearer to the point in time from which we may first return to the past. Though these currently insubstantial forms of evidence would also suddenly become more concrete if at any point this gap between the future and past is bridged by those who remain the only ones capable of doing so.

3.4.1. Diffident Descendants

It is understandable why our time traveling progeny would not wish to overtly interject themselves into a multitude of periods throughout their own past, particularly once the means with which to record such incursions had been developed. For instance, even though our distant descendants do not have to worry about "changing" history, for reasons that will be discussed in subsequent chapters, they would be incentivized to and might be legally required to limit explicit intertemporal contact, in order to avoid "adding" causal complexity in association with becoming a part of their own past.

Studying *Homo erectus*, the Nazca culture, or an isolated tribe in Sub-Saharan Africa in the distant past would leave little enduring evidence of contact, certainly as compared to plopping down in the center of Time Square in New York City during a New Year's Eve celebration, for example. Thousands of cell phone videos capturing the latter event would undoubtedly carry far more impact than an obscure cave painting or some geoglyphs in the desert of Peru. With this in mind, it is conceivable that we lack hard video evidence of recent close encounters, not because we are less interesting or worthy of study, but because stealthier measures are employed to investigate our period of the past. Larger population sizes, a more acute awareness of science and engineering, and the fact that a majority of people now carry mobile phones capable of taking pictures and videos, would make it harder to get in and out without being seen, and without leaving a permanent record of contact among people who lack an understanding of who extratempestrials are or why they are here.

If the avoidance of chrono-connective complexity is in the time travel code of our distant descendants, it could help explain why evidence of their visitation is so equivocal. However, it may still be possible for us now, as members of this more passive past, to garner some knowledge of that future state, from the testaments of those who have witnessed an IFO or extratempestrial in some capacity. Fortunately, there are a tremendous number of such accounts, including vetted reports provided by scientists, government officials, police, and military personnel, which also include details of a strange and reliable IFO event that may have happened entirely by accident.

3.4.2. Alleged Crash of an IFO at Roswell, New Mexico, U.S.A. 1947

An exemplary case of a well-documented IFO incident occurred near Roswell, New Mexico, in 1947. This occurrence was documented in a number of different ways, which included photographs, numerous eye-witness accounts, video of an autopsy performed on an extratempestri-al recovered from the crash site, newspaper reports, and even early ac-knowledgement by U.S. government officials that a flying disc had indeed crashed in the desert just outside this small town.

According to Kevin Randle and David Schmitt, authors of *The Truth about the UFO Crash at Roswell*,[70] the military had been tracking an uniden-tified flying object on radar for about four days leading up to the night of July 7, 1947, when this crash allegedly occurred. The following morning, the Roswell Daily Record ran a story stating that a flying disk had been recovered in the desert about 30–40 miles northwest of Roswell (figure 3-7). However, this story was soon retracted and replaced by a new ac-count in which U.S. government officials, while acknowledging that a crash did occur, stated that it was an experimental high-altitude surveillance balloon from a classified program named Project Mogul.

Later investigations launched by Stanton Friedman and Major Jesse Marcel—the latter of whom claims to have been involved in the initial recovery of the craft—brought to light evidence that this was indeed an event involving a crashed "alien" craft. Additionally, in 1989, mortician Glenn Dennis became the first witness to assert that he had seen distinctly human-like aliens at the Roswell Army Air Field and even provided a personal account of autopsies that were performed on them at that time.[70]

The United States Congress and Air Force eventually conducted inves-tigations into the Roswell incident, which produced two separate reports. The first stated that the debris recovered from the crash site was from the aforementioned surveillance program, Project Mogul. The second report, which was meant to specifically address the claim that alien bodies had been recovered from the crash site, alleged that these were simply crash test dummies used to test the surveillance balloon. Additionally, it stated that the dummies were confused for aliens because witnesses had distorted memories of what was perceived to have been a complex and traumatic event.[71]

Figure 3-7. *Initial publication in the Roswell Daily Record on July 8th, 1947 stating that a Flying Saucer had crashed near Roswell, New Mexico, U.S.A.*

Fifty years after this crash, Jim Wilson, Science and Technology Editor for Popular Mechanics, conducted an exhaustive assessment of historical data accumulated over the prior half-century. After completing this comprehensive survey, he concluded that an alien craft had indeed crashed in the desert outside of Roswell, New Mexico, on that night in 1947.

> After interviewing witnesses who had seen and handled crash-site debris and reviewing documents that were still classified when the GAO [U.S. Government Accountability Office] undertook its investigation, we have concluded that there really was a crashed disc, dead bodies and a secret that could have been politically deadly to presidents Harry S. Truman and Dwight D. Eisenhower.[72]

Few other IFO events have been examined as thoroughly as that of Roswell, and few approach the same level on Hynek's strangeness and probability rating scale as this incident in 1947. A major reason for this has to do with the corroborated claim that quintessentially human-looking individuals were recovered from the crash site and autopsied. If this is true, it would undoubtedly rank as one of the most monumental events in past-future human history. The implications of an incident such as this are vast, given that a tremendous amount of knowledge regarding our future biological, cultural, and technological state could be gleaned simply from examining what was recovered during this one single event.

We are currently unable to travel forward in time to glimpse these later stages of our biocultural evolution, but we would instantly become far more knowledgeable about that future if something from it were to come crashing into the past, as immeasurable details of this distant reality would suddenly become available to us.

Although I do not adhere to conspiracy theories, it is understandable why the U.S. government would be incentivized to limit public knowledge of a crash involving a much more sophisticated craft. If any nation were lucky enough to acquire a technologically advanced flying machine, it would undoubtedly allow them to jump far ahead in the millennia-old arms race and enduring battle for regional and now global military dominance. New technologies result from slight modifications to those that came before. However, working to reverse engineer a device that is the result of a much longer period of descent with modification could potentially catapult those who were fortunate enough to obtain it into a new and solitary realm of technological supremacy. This realization certainly must not have been lost on those who were tasked with recovering such a spectacle of human engineering in the New Mexico desert.

3.4.3. Foo Fighters and Close Encounters during WWII

The recovery of an extratempestrial craft at Roswell in 1947 may have ushered in a new phase of research and development for military machinery, though this was not the U.S.'s first close encounter. In fact, military pilots had often reported seeing large and unusual lights and aircraft in the skies above Europe and the Pacific during World War II. Dubbed "Foo Fighters," these craft were described as being fast and agile, performing aeronautic maneuvers that were not possible with any aircraft from that time period.

Reports of Foo Fighters were also very common and were often made by a number of different people at the same time, including whole crews aboard aircraft and naval ships after sightings were reported back by pilots in the air. Foo Fighter sightings were so common in fact that the Robertson Panel was asked to scrutinize this phenomenon specifically. Interestingly, while the panel concluded that these were not a direct threat to national security, they noted that Foo Fighters would be considered "flying saucers," had that term been in use at the time.

> Instances of "Foo Fighters" were cited. These were unexplained phenomena sighted by aircraft pilots during World War II in both European and Far East theaters of operation wherein "balls of light"

would fly near or with the aircraft and maneuver rapidly. … If the term "flying saucers" had been popular in 1943–1945, these objects would have been so labeled.[73][74]

Most nations have a propensity to romanticize instances of past warfare, and few capture our collective morbid interest to the same extent as World War II. Even after more than seventy years, there remains a widespread and somewhat sadistic fascination with it. If WWII continues to rank among the most interesting of international conflicts long into the future, it may help explain the unusually high level of IFO activity in and around regions most heavily engaged in this conflict throughout the mid-20th century, as it may become an important stop on any time tour of historical human conflicts.

3.4.4. The Gormon Dogfight—Fargo, North Dakota, U.S.A. 1948

The "Gormon Dogfight" was another aerial close encounter that occurred over the skies of Fargo, North Dakota, in 1948.[75] During this occasion, a World War II veteran named George Gorman pursued an IFO in his personal aircraft. After spotting the IFO, he engaged in what was deemed a "dogfight," despite the fact that Gorman was piloting a non-combative P-51 Mustang civilian aircraft, which was far less agile than the IFO he was shadowing. Because this encounter took place over U.S. airspace, occurred during a contentious time in international politics, involved an in-flight pursuit, and was observed concurrently by a number of different military personnel on the ground and in the sky, government officials were called to investigate. After examining the events associated with this encounter, they concluded that this high-speed pursuit of an IFO by a veteran Air Force pilot was simply an illusion involving a weather balloon.[75] In addition to undermining the credibility and integrity of George Gorman and his military service, this conclusion suggests that an epidemic of weather balloons had gripped the nation during the mid-20th century as this was, and would continue to be, the default explanation for IFOs for many years to come.

3.4.5. The Lubbock Lights—Lubbock, Texas, U.S.A. 1951

The Lubbock Lights involved an unusual formation of lights that were observed on multiple occasions flying over the city of Lubbock, Texas, in 1951. These formations were observed by numerous police, civilians, and perhaps most notably, by a group of professors from what is now Texas Tech University, whose advanced degrees and master status helped lend credence to their accounts.

According to their report, this group of professors was sitting outside in the backyard of one of their residences on the night of August 25, 1951, when they suddenly observed 20–30 lights fly overhead. While steeped in discourse about what they had just seen, a second group of lights in a similar formation reportedly flew directly over them once again. Later, on the night of September 5, 1951, these three professors, along with two other colleagues from Texas Tech, returned to the same private residence where they had originally observed this odd phenomenon, and again, witnessed this same light formation pass overhead of what was now a gaggle of academic observers.[76]

Noting that these objects were flying just above a group of thin clouds at about 610 meters off the ground, and based on how long the lights were in their field of vision, this group deduced that the craft or crafts were moving at approximately 600 miles per hour.[75] Lieutenant Edward J. Ruppelt, head of Project Blue Book at the time these sightings occurred, took up the task of examining these reports, calculations, and a group of photographs that were captured by a student at Texas Tech named Carl Hart. Based on the results of his investigation, Ruppelt determined that the Lubbock Lights were the result of recently installed vapor street lights reflecting off of a type of migratory bird called a Plover, which is common to that region of Texas.[75]

Naturally, the reflective Plover explanation was met with stark criticism by those who had observed and photographed this phenomenon, particularly considering that Ruppelt was attempting to gaslight them using actual gas lights. Additionally, when pictures were taken of real Plovers flying above these same vapor street lights at night, they in no way resembled the images captured by Carl Hart during the initial close encounter. Furthermore, according to Dr. Cross, a game warden and head of the Biology department at Texas Tech, the Plover hypothesis was highly unlikely because Plovers travel in much smaller groups than the number of lights observed, and because Plovers fly at relatively slow speeds, well below the 600-mph velocity estimated by these Texas Tech professors.[75]

The Lubbock Lights garnered ample attention from the media, created a buzz among local residents, and acted to bolster interest in and discussions about the existence of IFOs throughout the nation. Multiple events like the Lubbock Lights would continue to occur all throughout the United States during the remainder of the 20th century. However, the sightings that took place in the skies above Lubbock, Texas in 1951 remain to this day one of the most widely recognized and highly publicized events in IFO history.

3.4.6. U.S. Air Force Lieutenants Moncla and Wilson—Upper Peninsula of Michigan, U.S.A. 1953

Two years after the Lubbock Lights, another well-documented event that again involved U.S. military personnel took place on the Upper Peninsula of Michigan. On the evening of November 23, 1953, ground intercept radar operators in Sault Ste. Marie, Michigan identified an unusual blip on their radar screen located near Soo Locks on the St. Marys River. First Lieutenant Felix E. Moncla and Second Lieutenant Robert L. Wilson—the pilot and radar operator of an F-89 Scorpion—were dispatched from the Kincheloe Air Force Base to investigate.[77]

The onboard radar of their F-89 Scorpion had difficulty tracking the object, so Moncla was guided toward the object by ground command at Kincheloe. Ground control watched as the two radar hits got closer together and eventually merged into one, at which point it was assumed that they had flown over or under the other craft. However, after the two blips merged on radar, Moncla's disappeared entirely. American and Canadian search and rescue parties were unable to find any wreckage in the area of the disappearance, and no civilians in the region reported ever seeing or hearing the aircraft.[78]

The U.S. Air Force investigated this disappearance and concluded that the unknown craft on radar was a Royal Canadian Air Force Dakota C-47 that flew off course while traveling from Winnipeg to Sudbury, Canada.[78] However, the Royal Canadian Air Force and the pilot of the C-47 aircraft in question both adamantly deny this claim. Sadly, Moncla and Wilson were never found, nor was any wreckage attributed to the F-89 aircraft they were piloting while investigating this unknown radar anomaly.

3.4.7. Alien Abduction—The Case of Betty and Barney Hill, New Hampshire, U.S.A. 1961

Many reports of close encounters describe a more intimate form of contact between modern and future humans. In fact, survey research suggests that the number of people who have been abducted by IFOs could potentially be as high as 5–6 percent of the overall population.[79] Studies done using surveys are prone to sample bias and other methodological issues, so this statistic should definitely be approached with caution. Though even discounting what is likely an inflated statistical result, accounts of alien abductions are quite common and comprise a substantial number of authentic IFO reports.

There is some variation regarding the details of each specific abduction case, although they also tend to follow the same basic narrative overall. Generally speaking, during an abduction the individual reports feeling immobilized, being taken to a craft where physical and psychological tests are performed, body tissues, fecal matter, and gametes are collected, some communication takes place in the abductee's native language, and lastly, the abducted individual is returned near to where they were originally taken, often with a sense of lost time and fuzzy memories.

One of the first well-known and highly publicized IFO abduction events, which is also rather representative of abduction cases more generally, took place in a rural area of New Hampshire in 1961. This incident involved Betty Hill, who was a social worker at the time, and her husband Barney Hill, who worked for the U.S. Postal Service and sat on the board of their local U.S. Civil Rights Commission. According to their testimony, the two were driving home from a vacation on the night of September 19, 1961, when Betty glimpsed something that she initially thought was a shooting star. However, this bright light flew across the sky more like an airplane, rather than fading out like a shooting star.

Barney stopped the car to observe the light through his binoculars and initially concluded that it was an airplane, at least until the craft abruptly changed course and started heading toward them.[80] Somewhat alarmed, Barney and Betty began driving, but the craft caught up to them and rapidly descended from the sky ahead of their car, forcing Barney to stop in the middle of the road. According to Barney, the hovering craft looked like a huge pancake, and through his binoculars he could see 8 to 11 humanoid figures looking out at him from inside the IFO. All of these individuals except for one then moved out of view, and the one who remained began communicating with Barney, in English, telling him to "keep looking and that no harm would be done to you."[80] A couple years later while under hypnosis, Barney provided more detail about what ensued after he had stopped the car that evening:

> As he watched the leader through the binoculars, the leader's large eyes burned hypnotically into his mind and a "voice" within instructed him to keep coming closer, keep the binoculars to his eyes, and no harm would come to him ... he forced himself to let go of the binoculars, which dropped, breaking the strap around his neck.

As he hurried back to the car, the eyes and the mind-voice seemed
to follow, informing him further instructions would be issued and he
was not to tell his wife. Under hypnosis, Mr. Hill sketched the leader's
face, which showed enormous, slanted, oriental-like eyes, the peaked
cap, and something like a scarf over his left shoulder.[80]

After running back to the car, Barney remembered hearing a loud buzzing sound that filled the inside of their vehicle, which put the couple into
a trance-like state of dulled perception. A series of beeps later returned
them to full consciousness, at which point Betty and Barney realized that
they were 35 miles from where they had initially stopped on the road.
Both of them also possessed a sense that something extraordinary had
just happened, but they were unable to recall the details of what had
occurred at that time.[80] After returning home, they each took a long
shower and attempted to reconstruct the events of that night, aware that
something strange had just happened and that they were missing time and
many memories associated with it. Barney reported having had a strong
urge to examine his genitals, and Betty noticed that her dress shoes were
scuffed, her dress was torn, and there was an unexplainable pink powder
on it. The next morning, they also noticed small concentric circles on
the trunk of their car that had not been there previously, and when they
put a compass near the circles it spun vigorously but returned to normal
when moved away from the car.[81]

Betty began having vivid dreams about the encounter, which tended
to focus on the individuals that had abducted them. From her conscious
and subconscious memory, she described them as humanoid, standing
approximately five and a half feet tall, with bald heads, large eyes, small
ears, small noses, and greyish colored skin.[82] Betty also described the craft
as disc-shaped and metallic in appearance and recalled that while in a
trance-like state, both her and Barney were taken inside, separated, and
told they were going to be examined. While inside the IFO, Betty recalled
that a different "man" entered the room and performed an examination.
This individual, who she referred to as "the examiner," had a calm demeanor and spoke to her in English. However, she recalled that he did not
have a complete mastery of the language, as it was somewhat broken and
harder for her to understand compared to the other individuals aboard.
The examiner told Betty he was going to conduct a few tests to discover
differences between her and them. He then reportedly examined her eyes,
ears, mouth, throat, arms, legs, and hands, and also took some fingernail
trimmings and skin scrapings from her leg.[82]

Hoping that it would help reveal additional details about their abduction, Betty and Barney traveled to Boston, Massachusetts, to undergo hypnosis. A professional hypnotherapist named Dr. Benjamin Simon conducted separate sessions with Barney and Betty, and found that they provided similar accounts, with only slight differences in their recollection of events. Despite this high level of consistency between their independently provided reports, Dr. Simon concluded that Barney's abduction story was simply a fantasy created from Betty's dreams. However, as the Hills and others have pointed out, each of them separately recounted distinct details of the incident, indicating that Barney's unique experiences could not have been derived from Betty's account. Nonetheless, the Hills did manage to achieve some sense of closure from these sessions and all involved saw them as productive.[82]

The abduction of Betty and Barney Hill, as well as other cases like it, could potentially reveal certain details about the future of humanity. For if these are indeed our distant descendants, tasked with conducting anthropological research into the deep past from their own vantage point in the deep future, then we may be offered the opportunity to learn something of that future, while occupying the role of past research subject. That is, as long as we are willing to candidly consider reports provided by those who have experienced what is perhaps the closest form of a close encounter that anyone is capable of having.

3.5. Cross-Temporal Contact in Contemporary Times

There has been a relatively persistent rise in the number of IFO reports since the flurry of activity that characterized the mid-20th century. In fact, a simple query of the National UFO Reporting Center (NUFORC) database, which is the primary source for recording IFO and extraterrestrial encounters in the U.S., demonstrates just how common sightings have become in recent times.[83] For instance, between January 1973 and July 2014, a total of 88,386 reports have been filed, which works out to be about 177 sightings per month. However, these are not distributed evenly across this 41-year time period, but rather, there has been an accelerated rate of IFO reporting that began around the mid-1990s (figure 3-8).[84]

The discernible increase in the rate of reports over the last 20 years is interesting, and it could be interpreted in a few different ways. For example, this upward trend may be associated with a larger and continually increasing population size and therefore a greater likelihood that an event would be witnessed by someone. The timing of this uptick also correlates with the advent and more widespread use of the internet, which has made it easier to report IFO encounters. It is also possible that there has just been an increase in the frequency of IFO events since this time, although each of these is likely a co-factor in this broader relationship between sightings and time, to some extent.

The Canadian UFO Survey, which began documenting and detailing IFO events throughout Canada in 1989, is another useful source for understanding the frequency with which close encounters have occurred across North America in recent times.[85] It is also a valuable resource because this organization incorporates Dr. J. Allen Hynek's specific process for ascertaining the fidelity of any given report. More specifically, it makes use of his Strangeness and Reliability Scales and takes note of the number of witnesses of each encounter, which Hynek considered to be an important variable in assessing the likelihood and accuracy of any reported event.

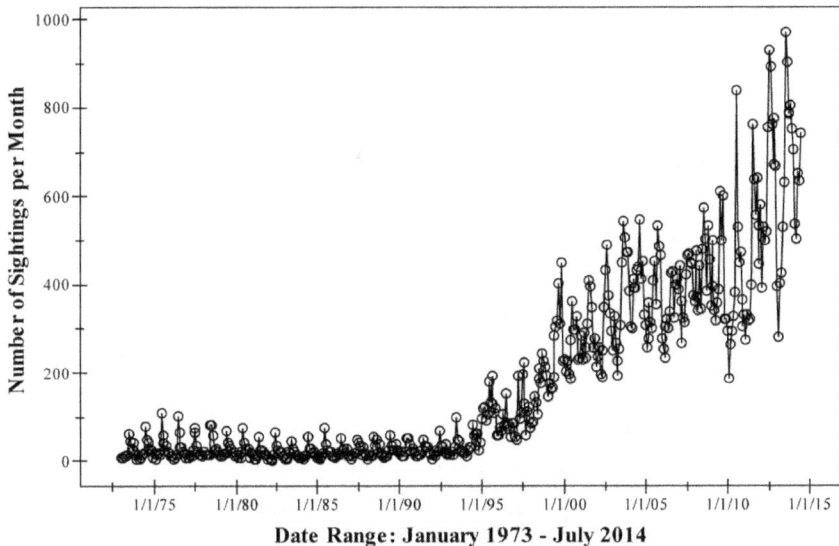

Date Range: January 1973 - July 2014
* Graph generated from data obtained from the National UFO Reporting Center (NUFORC) database

Figure 3-8. *Monthly statistics for close encounters reported to NUFORC, January 1973 –July 2014.*

A query of the Canadian UFO Survey database shows that 3,572 encounters involving IFOs and/or extratempestrials were reported over a 25-year period between 1989 (when they began collecting data), and 2014 (when this analysis was carried out).[86] Although these were recorded over a shorter timeframe and in a less populated country, the number of vetted sightings logged by the Canadian UFO Survey since 1989 is still quite large. This is particularly true considering that this is likely a more conservative estimate, given that the organization employs Hynek's rigorous criteria for registering IFO reports.

NUFORC does not make use of the same Strangeness and Reliability Scale, however these data are also reliable, as they scrutinize the reports provided to them using similar standards. For instance, NUFORC often calls witnesses to personally interview them and ask questions about their experience. This allows them to obtain more details regarding the specific event and helps assess the legitimacy of the report. Additionally, NUFORC identifies when a report is filed by police or military personnel, and they make it clear on the reporting form that "hoax" and "joke" reports will be ignored and immediately discarded.[87]

3.5.1. Modern Mindfulness of Close Encounters

NUFORC, the Canadian UFO Survey, and other institutions like these, represent a valuable resource for those interested in understanding people's experiences with IFOs. They also help us appreciate the frequency with which IFO events occur throughout North America, which is not likely to be known otherwise, given that people do not tend to speak openly about such encounters. In spite of numerous indications of contact throughout human history and prehistory and an increased rate of close encounters more recently—including some that occasionally involve hundreds or even thousands of people witnessing the same event at the same time, such as with the Phoenix Lights in 1997, for example[88]—widespread skepticism is likely to persist for some time.

There is no denying that any actual event involving IFOs falls far outside the established norms of society. Although, many of these seemingly odd encounters can also certainly be explained as conventional objects, or some type of everyday occurrence that was misconstrued by the observer. Some people may wholeheartedly believe that they have seen an IFO, when in fact it was actually just a typical aircraft, satellite, aerial research project, birds, hoaxes, hallucinations, stars, shooting stars, rising gas clouds, a SpaceX rocket returning to Earth, or even weather balloons. After all, the National Weather Service releases a staggering 75,000 radiosondes tethered to large and somewhat IFO-looking hydrogen or helium-filled weather balloons each year, all throughout the United States.[89]

It is easy to dismiss reports of unidentified aerial phenomenon, and we have certainly been culturally conditioned to do so. However, it seems unlikely that all IFO accounts can simply be explained away as something more typical of our time, particularly when they involve detailed accounts of abductions and anthropomorphic examinations. If taken seriously, descriptions provided by those who may have caught a glimpse of what's to come could potentially reveal much about the morphology, behavior, technology, language, and culture of our distant descendants, while also contributing to a clearer understanding of the otherwise opaque biocultural condition of humanity throughout the entirety of hominin evolution—past, present, and future.

4

Fermi's Paradox, Astrobiology, and the Question of Humanoid Extraterrestrial Life

The sure lesson of evolution is that organisms elsewhere must have separate evolutionary pathways; that their chemistry and biology and very likely their social organization will be profoundly dissimilar to anything on Earth.[90]

– Carl Sagan

4.1. The Search for Extraterrestrial Life

The idea that we may someday unravel the mysteries of time to the extent that our distant descendants become able to travel backward through it certainly seems like obscure science fiction. This is particularly true considering our relatively slow pace of progress in developing the materials, technology, and knowledge required to achieve backward time travel. Though for context, the notion that we would ever be able to hurl ourselves off of planet Earth and onto the moon or that we could send research vessels to Mars, Pluto, or to intercept a fast-moving comet must have also seemed outlandish throughout the vast majority of human history.

Compared to time, we have learned much more about space, even venturing out into our proximate celestial neighborhood to some extent. Although time and three-dimensional space are actually fused into a single four-dimensional continuum, aptly dubbed spacetime, we still tend to separate them in a more colloquial sense. As such, it is understandable why the dominant worldview regarding alien encounters has thus far centered on spacefaring extraterrestrials, without much thought given to the possibility of human interaction across different periods on this planet.

The notion of contact and visitation by intelligent extraterrestrial humanoid life outside our solar system is riddled with hitches, which will be discussed in detail throughout these next two chapters. With that said, there is no doubt that some form of life exists on other planets, and particularly considering how quickly it happened here on Earth once our rocky planet became capable of sustaining life. The search for these simpler forms of life is a primary focus of astrobiologists, who investigate whether other habitable planets exist and whether other lifeforms live or have lived upon them. However, the issue of advanced humanoid life that happens to exist close enough to us and at the same time so that communication and contact would be possible is a fundamentally different question.

In the 1950s, physicist Enrico Fermi pointed out that our sun is relatively young and that it is similar to other stars at the center of other solar systems. Because there are billions of these older stars in the universe, which are also likely to have Earth-like planets orbiting them, Fermi speculated that intelligent life capable of interstellar travel must have developed on a certain percentage of them. However, this realization sparked a philosophical quandary that came to be known as Fermi's Paradox, which simply asks the following: if there is such a high probability of advanced extraterrestrial civilizations in the universe, then where is everybody?[91]

After the U.S. developed nuclear technology in the mid-1900s, Fermi began to posit that other civilizations on other planets must have also discovered the same universal laws of the physical world that allow for this technology. Though he wondered why we still lack evidence of these civilizations, since he expected that they would eventually expand into our region of the universe.[92] However, other physicists were not so optimistic. In fact, famed physicist Frank Tipler even went as far as to publish a paper in the journal of the Royal Astronomical Society, with the somewhat audacious title *Extraterrestrial Intelligent Beings Do Not Exist.*[93]

Regardless of its complexity, the National Aeronautics and Space Administration (NASA) is confident that we will discover some form of extraterrestrial life relatively soon. In fact, in a recent public panel discussion in Washington D.C., NASA chief scientist Ellen Stofan stated "I think we're going to have strong indications of life beyond Earth within a decade, and I think we're going to have definitive evidence within 20 to 30 years. We know where to look. We know how to look. In most cases, we have the technology, and we're on a path to implementing it."[94]

Presumably, most extraterrestrial lifeforms do not beam radio signals into space, which is indicated by the fact that we remain the only organism capable of doing so in the entire 3.7-billion-year history of life on Earth. Instead, astrobiologists scouring the surface of Mars, Comet 67P, and planets outside our solar system, are primarily searching for *biosignatures*—things such as excess oxygen in the atmosphere, water, hydrogen, carbon, and other basic components of complex organic molecules that have been vital to the formation of life on Earth.

Because simple lifeforms are small, because immense distances separate us and other planets, and because of interference from light emissions coming from the sun and other stars, astrobiologists are not able to simply zoom in with high-powered telescopes to look for plants and animals on the surface of distant planets. As a result, they primarily make use of spectroscopy, which examines light emission and absorption patterns at various wavelengths as it interacts with different forms of matter, on what is a rapidly expanding list of exoplanets.[95]

Complex lifeforms inevitably begin as simple ones, growing more intricate with incremental cumulative changes, over what is to us at least, a very long time. You don't just get tapirs, fire ants, and wallabies without simple unicellular life first developing from basic molecules in warm pools of water. But what happens after the early formation and replication of life is completely dictated by the specific environmental conditions on each specific planet. The amount and ratio of different elements, the planet's size, force of gravity, distance from its sun, temperature, ability to sustain an atmosphere and liquid water, as well as a multitude of other factors, are all important to the development and maintenance of life over the long term.

Surely, some other forms of life exist on other planets elsewhere in the vastness of space. However, we must consider more intricate details regarding what they might look like, whether they are intelligent, and whether they exist in an advanced state at the same time as us so that we may discover each other living around our respective stars. It is also important to skeptically consider other problems associated with traversing titanic reaches of the cosmos to visit each other, and issues related to how we would communicate with creatures that have undergone a separate evolutionary process under what must have been markedly different planetary, environmental, and social conditions.[96][97][98]

4.2. The Drake Equation and Probability of Humanoid Life on Earth-like Exoplanets

The Drake Equation represents one of the earliest attempts to address the more general question of the probability of intelligent life on a separate planet elsewhere in the Milky Way galaxy. The equation was initially developed by astronomer Frank Drake and was meant to stimulate discussion at an informal gathering of the Space Science Board-National Academy of Sciences conference at the Green Bank radio telescope facility in West Virginia in 1961.[99][100] History would show that the Drake Equation was very successful in this regard, as it quickly became a standard jumping-off point for attempts at estimating the number of advanced civilizations in our galaxy that may be capable of interplanetary communication, which was a variable designated as (N). A number of astronomical, technological, and cultural variables were factored into the equation. These included such things as the rate of star formation per year (R^*), the percentage of stars that may have planets in orbit around them (f_p), the number of those with habitable planets (n_e), the fraction of those that develop biological life (f_l), the fraction of those that develop intelligent life (f_i), and those that are willing and able to communicate beyond their terrestrial borders (f_c).[100]

Of equal importance to the question of "do they exist," is the question "would it be possible to find and communicate with them while they are in existence." This question gave rise to the final variable in the equation (L), which considers the amount of time that a civilization may be producing a detectable signal.

In total, Drake's equation reads: $\mathbf{N = R^* \cdot f_p \cdot n_e \cdot f_l \cdot f_i \cdot f_c \cdot L}$

Variable	Description
N	The number of civilizations in The Milky Way Galaxy whose electromagnetic emissions are detectable.
R^*	The rate of formation of stars suitable for the development of intelligent life.
f_p	The fraction of those stars with planetary systems.
n_e	The number of planets, per solar system, with an environment suitable for life.
f_l	The fraction of suitable planets on which life actually appears.

Variable	Description
f_i	The fraction of life bearing planets on which intelligent life emerges.
f_c	The fraction of civilizations that develop technology releasing detectable signs of their existence into space.
L	The length of time such civilizations release detectable signals into space.[100]

In 1961, Frank Drake and his collaborators attempted to assign actual numerical values for each of these variables, which were as follows:

Variable	Value	Description
R^*	1/year	1 star formed per year, considered a conservative
f_p	0.2–0.5	25–50% of all stars formed will have planets
n_e	1–5	Stars with planets have 1 to 5 planets capable of supporting life
f_l	1	100% of these planets will develop life
f_i	1	100% of which will develop intelligent life
f_c	0.1–0.2	10–20% of these can carry out interstellar communication
L	1 K – 100 M years	These civilizations will last between $1,000 - 100,000,000$ years.[101]

Using the lower value of each range given for each variable in this equation produces a value of N that is approximately 20. Using the higher value in each range produces an estimate of N that is closer to 50 million. In other words, Drake and his colleagues estimated that the number of advanced civilizations in our galaxy that may be capable of interplanetary communication is somewhere between 20 and 50,000,000.

If history has taught us anything, it is that civilizations come and go. Throughout this historic ebb and flow of the rise and fall of societies, we have slowly moved toward becoming one large highly integrated global population. This could potentially make us susceptible to a much larger collapse, though in the meantime, various nations have joined forces to develop the types of space projects that would get us noticed to an extra-terrestrial civilization. However, in the grand scheme of our 6-million-year hominin history, we have only been doing detectable things for a very small portion of this time.

The first human shot into space from our planet was Youri Alekseyevich Gagarin, which occurred on April 12, 1961. The first intentional electro-magnetic signal emitted from Earth was on November 16, 1974. These are very recent advances in the context of the long history of hominins, and the even longer history of life on Earth. This is important in the context of Frank Drake's (f_c) and (L) variables, where a civilization, including our own, must develop the right technology at just the right time to be noticed by another civilization who has done the same. This makes for a short contact window between the time that the technology arises and when the civilization implodes or evolves beyond the phase of recognizable technology. According to Milan Cirkovic, author of the 2004 *Astrobiology* article titled "The temporal aspect of the Drake equation and SETI":

> Thus, the set of the civilizations interesting from the point of view of SETI is not open in the temporal sense, but instead forms a "communication window", which begins at the moment the required technology is developed (factor f_c in the Drake equation) and is terminated either through extinction of the civilization or through it passing into the realm of "supercivilizations" unreachable by our primitive SETI means.[102]

It doesn't take an astronomer, astrobiologist, anthropologist, or statis-tician to realize that the majority of values used in the Drake equation are speculative guesses. This is also supported by the fact that Drake's original estimate for the number of intelligent civilizations in our galaxy ranges from 20 all the way up to 50 million. Though as our technology and knowledge of the cosmos continue to evolve, it is likely that we will be able to revise some of the estimates used in this equation. In fact, as of May 2013, the Kepler Mission—tasked with searching for planets outside our solar system—had confirmed 978 planets in 400 different star systems, while also identifying an additional 4,234 potential planetary candidates.[95]

This is a rather impressive number, considering it had only been about 25 years since 51 Pegasi b—the first true planet orbiting a sun-like star outside our solar system—was discovered on October 6th, 1995.[103] While this is encouraging, at least with respect to Drake's (f_p) and (n_e) variables, the majority of estimates for the other factors in the equation are likely to remain more speculative. This is because Earth is the only data point in our very limited sample of habitable planets that have produced and sustained life, and it is impractical to extrapolate our condition to other planets in the galaxy and to the universe beyond, from a sample size of one.

Additionally, the probability that any condition in the equation would be met entirely is so low that together the number of advanced civilizations in our galaxy must be much smaller than 50 million and likely even smaller than the low estimate of 20. On what grounds can we claim that 100% of planets that can develop life will do so (f_l) or that 100% of those that develop life will evolve intelligent life (f_i)? Likewise, how can we assume that 10–20% of those that develop intelligent life will be able to communicate beyond their star, especially considering how incredibly rare this ability is among lifeforms and time periods on our own planet?

4.2.1. *Terrestrial and Extraterrestrial Physiology*

Other scientists are also skeptical about the universal ubiquity of intelligent life on other planets (f_i) or that extraterrestrial beings would ever evolve to look anything like us—assuming they do exist somewhere else in the universe. The quote by Carl Sagan that began this chapter cautions that extraterrestrial organisms must have separate evolutionary pathways, which will result in profoundly dissimilar organisms to anything on Earth. This sentiment is also echoed by anthropologist Garry Chick in the article *Biocultural Prerequisites for the Development of Interstellar Communication*:

> Somewhere between 1.5 and 2 million living species have been cataloged on Earth....When 20 species of hominoids are included in that total of 2 million extant species, primates constitute only 0.001 percent (0.00001) of the living species on Earth. Moreover, only 1 of these 20 species has developed a technology capable of interstellar communication. In sum, the development of high intelligence on Earth has been extremely rare, and there is little evidence to support the idea that its development is inevitable. Even if some forms of intelligence do evolve on other planets, there is no good reason to believe that at least one of them must be human-like. Hence, high estimates of fi may be not only anthropocentric, but also highly optimistic.[104]

Additionally, according to Douglas Vakoch, president of Messaging Extraterrestrial Intelligence (METI) and editor of the NASA-sponsored research paper *Archaeology, Anthropology and Interstellar Communication*:

> Moreover, any civilization we contact will have arisen independently of life on Earth, in the habitable zone of a star stable enough to allow its inhabitants to evolve biologically, culturally, and technologically. The evolutionary path followed by extraterrestrial intelligence will no doubt diverge in significant ways from the one traveled by humans over the course of our history.[105]

It is currently impossible to know how many times life arose on other planets. It is especially difficult to know how many times encephalized, bipedal life that blindly projects radio waves and communication probes out into space arose around other stars in this galaxy. Bipedalism and encephalization—an evolutionary increase in the relative size and complexity of the brain—are particularly noteworthy because these two factors were instrumental in the development of our own unique form of intelligence. Bipedalism and encephalization are also important to consider because reports of close encounters with alien beings ubiquitously describe them as big-brained, small-faced, large-eyed, bipedal humanoids, which, as indicated by Carl Sagan and Gary Chick, is a physiological form that is not likely to evolve on a separate planet elsewhere in the universe. In fact, the best available data suggest that the number of times a species has evolved these specific morphological characteristics, anywhere and in the entire history of the universe, is one.

Bipedalism and encephalization are certainly not prerequisites for intelligence, on this or likely any other planet. Indeed, some of the smartest organisms on Earth are not bipedal (i.e., elephants, canids, dolphins), or encephalized (octopi, ravens, etc.). However, considering the evolution of our species—who exist as the sole animal capable of contemplating and researching our own existence—our large brains and free use of our hands are largely the result of our bipedal form of locomotion.

Upright standing and walking, which arose among our hominin ancestors approximately 6–8 million years ago in eastern and northern Africa, set in motion a number of important morphological and behavioral changes. Most notably, standing upright created the need to reorient our field of vision downward so that we could continue to see where we were going and to see predators and other physical hazards approaching. For instance, if a chimpanzee were to stand up on two legs, the default orientation of its eyes would be upward toward the sky, given that they have evolved to walk quadrupedally, with their eyes oriented toward the horizon while doing so.

Standing upright meant that the visual plane of early hominins needed to reorient downward, which also involved a downward rotation of the entire face and head. To facilitate this change, and to help center our skulls upon what was becoming a vertically oriented spine, the foramen magnum (the big hole at the base of our crania) began to move anteriorly toward the front of our skulls. Together, these morphological changes allowed us to look forward, while also balancing our skulls upon our vertebrae. The combined result of this cranial and facial juxtapositioning was that more space became available toward the top of our skulls where we could begin to grow increasingly larger and more complex brains. In other words, the downward and backward rotation of the face, combined with the forward movement of the foramen magnum, resulted in a flexing of the entire cranial base, which is the group of bones that form the base of the cranium upon which the brain sits within our skulls. This process is known as cranial flexing and is an important part of understanding how and why humans have become so highly encephalized in association with our shift toward bipedalism.

A good way of conceptualizing these long-term craniofacial changes is to imagine bending two ends of a slinky downward and toward each other. As an analogy for the evolution of the hominin skull, the front end of the slinky represents the face, and the back end of the slinky the foramen magnum. As they move toward each other, the middle of the slinky begins to bulge upward and outward, creating a larger rounded area. This is similar to how our calvaria, or upper braincase, expanded upward and outward throughout hominin evolution, creating more space for our expanding and increasingly intelligent brain.

Bipedalism was also important because it freed up our hands to begin doing new and useful things, including making stone tools for instance. The first stone tools were simple, unifacially flaked, chunky pieces of relatively poor-quality stone from the site of Lomekwi 3 in West Turkana, Kenya, dating to approximately 3.3 million years ago.[106] In fact, these early stone tools are a direct ancestor of computers, cell phones, sporks, airplanes, and every other tool in existence today, including those that help us search for other advanced civilizations capable of interplanetary communication. These most inimitable characteristics of humanity have been shaped by a multitude of different environmental, ecological, and social forces over the last 6 million years of hominin evolution—and further forces throughout the last 3.7 billion years of life on Earth. Considering all of the factors that have gone into making us just the way we are today, it is incomprehensible how these same processes could ever play out in the exact same way, anywhere else in the known universe.

Bipedalism was undoubtedly an important factor contributing to our advanced state of intellect and culture, which we owe to the specific conditions that favored upright standing and walking in our ancestral homeland of Africa 6–8 million years ago. This form of locomotion, as well as the benefits it has imparted, may also be a result of the specific planetary characteristics of Earth, as we are seemingly toward the upper limit of any planet capable of sustaining terrestrial bipedal locomotion, given the relatively large size, mass, and therefore gravity of Earth. This is indicated by the fact that most earthly organisms walk on four or more legs, as well as the fact that we continue to suffer a host of problems stemming from our bipedal form of locomotion, despite having done it for over 6 million years now. In an editorial in the *Annals of Biomedical Engineering*, Dr. Bruce Latimer, an anthropologist and former director of the Cleveland Museum of Natural History, describes what he refers to as the *Perils of Being Bipedal:*[107]

> Indeed, it is unfortunate but true that if we live long enough, nearly all of us will suffer from being bipedal. Not just aching feet, sprained ankles, or arthritic knees and hips, but a whole host of conditions that are as unique to our species as is our peculiar way of walking. For example, only our species regularly endures such common maladies as fractured hips, bunions, hernias (inguinal and femoral), fallen arches, torn menisci, shin splints, herniated intervertebral discs, fractured vertebrae, spondylolysis, scoliosis, and kyphosis—just to name a few.[108]

The gravitational pull of Earth is 9.8 m/s². Had we evolved on a planet with a lower mass, we may have developed a form more conducive to hopping or floating. Alternatively, if we had evolved on a planet with greater mass and gravity, crawling, rolling, or not moving at all may have been preferable. The fact that we endure so many problems associated with walking upright on a planet with a gravitational force of only 9.8 m/s² indicates that bipedalism may be even rarer on planets larger than Earth, and, as it turns out, the vast majority of exoplanets are much larger than our own.

Based on data obtained from the Exoplanet Catalog of the Planetary Habitability Laboratory at the University of Puerto Rico, Arecibo, only

46 of the 2,023 confirmed exoplanets, or 2.27% of them, are estimated to have a mass less than or equivalent to that of Earth.[109] This shows that Earth is a relatively small planet and indicates that bipedalism, which is already rare here, may be even less common on other planets in the galaxy.

Issues related to how gravity and other planetary characteristics shape the form and function of organisms on different planets has been the subject of recent studies in astrobiology, biomechanics, and in the development of bio-suit systems for astronauts who may embark on future space and planetary exploration missions.[110 111] Questions related to exoplanetary gravity and the biomechanics of living organisms were also explored in a recent Public Broadcasting Service episode of NOVA, called *Alien Planets Revealed*.[112] This program highlighted the work of Dr. Bill Sellers, a biomechanics researcher and professor of Life Sciences at the University of Manchester in England. He uses computer simulations of animal movements, along with a "genetic algorithm" that selects for the most efficient form of locomotion out of hundreds of randomly generated movement patterns, to assess optimal body-supporting leg compositions on high-gravity planets. After applying this technique to a hypothetical alien creature on a high-gravity planet similar to that of Kepler-62f, which lies in the habitable zone of its sun and is approximately 1.41 times larger than the radius of Earth,[113] the result was an organism with a very stable and efficient eight-legged gait.[112] This research again points to the rarity of bipedalism evolving on a planet that is even slightly larger than our own and instead suggests that octopedalism might be a more common form of locomotion under these types of gravitational conditions.

Additionally, according to Dr. Lewis Dartnell, an astrobiology research fellow at the University of Leicester, creatures on more massive planets are likely to be grazers and possess large elephant-like trunks that allow them to feed without having to raise up and down against the stronger force of gravity.[112] In this way, high-gravity environments may not only influence the physiological form of organisms, but could play a part in shaping their behavioral characteristics as well. In fact, a behavioral response to changing gravity was recently demonstrated experimentally, when eight colonies of 80 ants (*Tetramorium caespitum*) were taken aboard the International Space Station. This research, published in a 2015 *Frontiers in Ecology and Evolution* paper titled *Collective Search by Ants in Microgravity*, examined how ants search new areas beyond their nesting site in zero gravity.[114] On Earth, this experiment showed that providing extra space for the ants reduced their overall density, as they altered their routes to spread out and cover more territory, to the extent that every corner of the box was visited by multiple ants within a five minute timeframe. However, in zero gravity, the ants were far less effective at searching these new areas, taking more convoluted paths around them, and leaving many parts unexplored.[115]

The myriad effects that gravity has on shaping the physical and behavioral characteristics of an organism, along with the fact that the vast majority of nearby planets are more massive than Earth, suggests that even if intelligent extraterrestrial beings do exist, we are likely the only intelligent bipedal beings in the galactic neighborhood. Furthermore, our search for habitable exoplanets continues to demonstrate the uniqueness of our own planet in size, orbit, atmosphere, gravity, and chemical composition, making it highly unlikely that an extraterrestrial being on a nearby planet would ever evolve to look, walk, eat, drink, sense, think, communicate, or generally be anything like us at all.

4.3. Interplanetary Evolutionary Convergence and the Double Coincidence of Time

The continual discovery of Earth-like exoplanets helps generate more realistic estimates that can be used in the Drake equation for approximating the likelihood of intelligent life capable of interstellar communication. Thus far, this search has demonstrated a high level of variability in the chemical properties and overall structure of planets and solar systems. This speaks to the relative uniqueness of Earth and to the way in which countless variables have come together in just the right way to make our big-brained bipedal species. However, not all researchers agree that our physical and cognitive state is as distinctive as it may seem. In his book *The Runes of Evolution*, Simon Conway Morris, a professor of evolutionary paleobiology at Cambridge University in England, advocates that all planets within the habitable zone of a star will inevitably develop intelligent life because of convergent evolution.[116] In a press release leading up to the publication of his book, Morris states:

> I would argue that in any habitable zone that doesn't boil or freeze, intelligent life is going to emerge, because intelligence is convergent. . . . One can say with reasonable confidence that the likelihood of something analogous to a human evolving is really pretty high. . . . And given the number of potential planets that we now have good reason to think exist, even if the dice only come up the right way every one in 100 throws, that still leads to a very large number of intelligences scattered around, that are likely to be similar to us.[117]

We have discovered a large number of planets as part of NASA's expansive Kepler Mission. However, NASA, SETI, and other organizations have not yet discovered any evidence of intelligent life on any of these planets, or anywhere else for that matter. If intelligent life were an inevitability on every planet in the habitable zone of its star, one would think that the heavens would be buzzing with radio signals and other electromagnetic indications of this intelligence. Additionally, the high degree of variability in planetary characteristics and the complexity of networks in a planet-specific evolutionary web of life make universal interplanetary evolutionary convergence a hard pill to swallow.

Natural selection works on individuals, but it is the population that evolves and adapts. This process is also entirely random, with countless factors shaping the characteristics of species in accordance with each unique environment over tremendously long periods. There are certainly examples of homoplasy (a.k.a. analogies, evolutionary convergence) here on Earth, in which unrelated animals evolve similar morphological forms, such as with the wing of bats, birds, and various insects, for example. These functional morphological traits evolved independently of one another, as they are not present in the common ancestor of any of these animals. However, a different planet, orbiting a different star, with a different temperature, climate, wind speed, gravity, atmosphere, chemical makeup, etc., would not be expected to directionally select for the same kinds of traits every time, and certainly not always toward something "analogous to a human."

Evolution has taken many different trajectories on Earth, as indicated by instances of camouflage, mimicry, homology, homoplasy, as well as mass extinctions and other periods of punctuated stasis. Major events, like the 65 million-year-old asteroid impact on our planet, can lead to widespread ecological change, including adaptive radiations, when organisms rapidly diversify into a multitude of new forms. Even minor events and subtle changes to the environment can set in motion a period of change in the physiology and behavior of various organisms, which remain specific to those environmental conditions. Constantly shifting landmasses, the Earth's distance from the sun, the wobble of the Earth around its axis, etc., have all contributed to changing climatic conditions and modification to local and global environments, resulting in stochastic shifts in the morphological and behavioral characteristics of lifeforms that lived within these planet-specific environments.

Regardless of how earthlike another planet may be, it seems utterly impossible that extraterrestrial life would ever follow the exact same path, throughout 3.7 billion years of evolutionary change, and result in something that looked anything like a human. Too many specific random occurrences went into making us just the way we are. In fact, I would argue that the probability of a fleshy, big-brained, bipedal, pentadactyl, highly intelligent lifeform arising independently on a different planet orbiting a nearby star is effectively zero, given the countless factors necessary to create and sustain this same evolutionary trajectory over such an extraordinary amount of time.

Instances of convergent evolution here on Earth are the result of similar problems resulting in similar solutions, in similar environments, and most notably, on the same planet. It is hard to imagine that any large brained, pentadactyl, bipedal tetrapod—which is unique even among the millions of species on this planet—would ever evolve again on a different planet. Let alone that it would be an inevitability on all planets in the habitable zone of all stars throughout the universe.

Fergus Simpson, a cosmologist at the University of Barcelona in Spain, used Bayesian statistical models along with planetary and population size data to predict what types of physical traits intelligent extraterrestrial lifeforms may have. The results of his research suggested that intelligent life on other planets would not only be very different from humans, but that it would be about five times larger than us, or about the size of a polar bear:

> Throughout the animal kingdom, species which are physically larger invariably possess a lower population density, possibly due to their enhanced energy demands. As a result, we should expect humans to be physically smaller than most other advanced species. By marginalising over a feasible range of standard deviations, we conclude that most species are expected to exceed 300 kg in body mass. The median body mass is similar to that of a polar bear.[118]

Further evidence that humanoid life is not ubiquitous on other planets comes from the fact that it has only evolved once on this planet. In other words, if intelligent humanoid beings were inevitable everywhere, then we would expect to have seen convergent evolution result in multiple different intelligent human species that do not share a common ancestor here on Earth. However, the fossil record paints a different picture, showing that traits common among extant and extinct hominins are the result of shared ancestry, rather than similar environments shaping similar features across unrelated taxa. Without evolutionary convergence leading to separate intelligent humanoid lifeforms on this planet, it is hard to imagine that this would be an inevitable outcome on all planets capable of sustaining life in the universe.

Additionally, even if two separate planetary systems were both able to produce similar humanoid lifeforms, it is highly unlikely that they would exist in the same stage of evolutionary development at the same time. According to a 2015 article titled *On the History and Future of Cosmic Planet Formation*, authors Peter Behroozi and Molly S. Peeples from the Space Telescope Science Institute in Baltimore, Maryland, note that Earth may have formed earlier than 92% of all other potentially habitable planets in the universe.[119] This means that we were quite early to the party relative to the vast majority of other planets like our own. This also implies that intelligent extraterrestrial life is rather uncommon in the universe, as it still lacks a planet on which to begin the long process of developing into a complex cognizant organism, if it is to happen at all. According to renowned physicist Frank Tipler, in the above-mentioned article *Extraterrestrial Intelligent Beings Do Not Exist:*

> The biologists argue that the number of evolutionary pathways leading from one-celled organisms to intelligent beings is miniscule when compared with the total number of evolutionary pathways, and thus even if we grant the existence of life on 10^9 to 10^{10} planets in our Galaxy, the probability that intelligence has arisen in our Galaxy on any planet but our own is still very small ... the probability of the evolution of creatures with the technological capability of interstellar communication within five billion years after the development of life on an Earth-like planet is less than 10^{-10} and thus we are the only intelligent species now existing in this Galaxy.[120]

He goes on to write about another important element of this debate that echoes Fermi's paradox, which is simply that if intelligent life is inevitable, or even common, then we surely would have already been visited by, or at least contacted by, one of the many intelligent civilizations out there:

> Extraterrestrial intelligent beings do not exist: if they did exist and possessed the technology for interstellar communication, they would also have developed interstellar travel and thus would already be present in our solar system. Since they are not here, it follows that they do not exist.[120]

The fact that life arose on planet Earth shortly after it became geologically and ecologically possible (i.e., dropping temperatures, a hardening crust, a sustained atmosphere, liquid water), indicates that it also could have happened on any number of other planets in the habitable zone of their star. In fact, based on the rapidity of biogenesis on Earth, it has been estimated that the probability of life arising on Earth-like planets older than one billion years is >13%.[121] However, there remain a number of additional variables—which the Drake equation attempts to account for—that make the probability of contact or visitation by an intelligent extraterrestrial lifeform highly unlikely.

One could argue that we are still too close to when we first began sending messages and unmanned probes into space to expect that they would have elicited a response by now. This is particularly noteworthy considering that it would take the same amount of time for a return signal to get here as it took our initial signal to get there (wherever there might be). Taking into account the tremendous amount of space and time between solar systems and the improbability that an organism on a distant planet will be at just the right place along its own evolution path to possess the intelligence and technology necessary to send and receive radio transmissions, it is likely to be a very long time before anyone hears anything, if at all.

Even if an advanced extraterrestrial civilization were to develop technology capable of transmitting and receiving messages and were able to locate us in the vast reaches of empty space, the issue of mutual understanding is still a potential roadblock to meaningful communication. Considering the multitude of languages and different forms of communication on our own planet, as well as our utter inability to carry on complex conversations with any other earthly animal, there is little evidence to support the idea that interstellar communication would be easy, or even possible. Furthermore, even if we were somehow able to decode an alien language, it is unlikely to ever result in mutual visitation, given all of the many logistical problems of interstellar travel and transportation.

5

Occam's Razor: The Enigma of Interstellar Space Travel, Contact, and Communication with Extraterrestrial Life

The distances between stars are immense. Consider Voyager 1, the fastest space-craft in the solar system. It took 18 months to reach Jupiter. Thirty-eight years after launch, it has fled 12 billion miles from the sun. At this rate, it will travel the distance to Alpha Centauri, the nearest star system, in about 70,000 years. Getting to Kepler 438b would take a boggling 8 million years, considerably longer than our species has been around.[122]

– Corey S. Powell

5.1. Space, Time, and Interstellar Travel

In a monumental 1905 paper titled *On The Electrodynamics of Moving Bodies*, Albert Einstein showed that the speed of light is a universal constant that governs the whole of four-dimensional spacetime.[123] An important implication of this early research is that people are capable of experiencing the passage of space and time differently, depending on their speed relative to others and to that of light. For instance, if someone were to board a spacecraft and travel near the speed of light, they would experience a shortening of both space and time in their frame of reference. In other words, they would see space contracting around them, and time would pass more slowly relative to those who stayed behind.

An integral part of Einstein's theory of special relativity is known as time dilation, in which the effect of high-speed motion on the passage of space and time is different for people in different inertial reference frames. This case is often depicted in the context of the "twin paradox," which is where one member of a twin pair sets off on a high-speed round-trip voyage and upon returning home, discovers that her/his (formerly) identical twin is now much older by comparison (figure 5-1). In spite of the name, this scenario is not actually paradoxical in any way, although it is a good metaphor for conceptualizing the relativity of space and time among individuals in different reference frames.

Figure 5-1. *The effect of time dilation is often depicted in the context of the "twin paradox," in which two identical twins experience the passage of time differently after one embarks on a high-speed voyage.*

Even though space and time become compressed while traveling at full tilt relative to the speed of light, the observer inside the rocket ship does not perceive time running more slowly. Clocks within the rocket continue to tick at the same rate, and the space traveler's biological perception of the passage of time remains the same. It is only upon returning to Earth and comparing clocks and wrinkles with those who stayed behind that the effects of this prolonged high-speed voyage are realized.

Rapid motion relative to the speed of light is also important to consider in the context of time travel, as time dilation in this capacity represents a simple form of time travel to the future. The speed limit of light is also important for understanding constraints associated with long-distance space travel, contact, and communication with lifeforms that may potentially be living on planets in different solar systems throughout the cosmos. For instance, imagine a creature looking at Earth through a very large telescope on a planet that is 67 million light-years away. Even though they are looking at Earth right now, they would actually still see dinosaurs rummaging around. Light is capable of traveling at 299,792 kilometers per second, however, this planet is about $6.34e^{20}$ kilometers away, meaning that it takes 67 million years for the light that is reflected off of Earth's surface to arrive at the lens of this being's telescope. Instead of seeing us and our bright lights and big cities as they exist "now," these extraterrestrials are only able to see what existed at that point in our past when light began its long journey toward them.

In this same context, everything we see while looking out into deep space is a snapshot in time of remote periods of the ancient celestial past, outlying gaseous and incandescent bodies with relative states of temporal ancestry defined by their distance from Earth. Because of the immense distances between solar systems and galaxies, which continue to move farther away from each other at high velocity, we are not just separated by a vast expanse of space, but also by incredible amounts of time.

In 1974, the 905-foot-tall antenna at the Arecibo Observatory in Puerto Rico broadcasted an encoded radio message out toward a globular cluster of stars known as Messier 13 (M13), which are approximately 25,000 light-years from Earth. The message was written by Frank Drake, Carl Sagan, and others, and consisted of 1,679 binary digits containing information about our numbering system, DNA, physical characteristics of humans, a description of our solar system, and information about the Arecibo radio antenna itself.[124]

Radio signals also travel at the speed of light. However, because M13 is 25,000 light-years away, this message, traveling at 299,792.46 kilometers per second, or about 186,164 miles per second, will still take 25,000 years to reach it. If there happens to be an intelligent lifeform on a planet in this star cluster that also uses radio communication technology, and they can figure out what we said in our message and how to respond to it using our symbolic language, we would not expect their reply to reach us until at least 51,974 A.D.

The speed of light imposes constraints on the rate with which we can engage in interstellar communication, but it becomes even more problematic when considering interstellar space travel. For instance, imagine that instead of sending a radio broadcast to M13 in 1974, we had actually sent it about 52,000 years ago, and just this year, we finally received an invitation from our distant neighbors to join them for tea. We recruit a group of brave astronauts to serve as Earth ambassadors, and we employ a team of top scientists and engineers to construct a craft capable of sustained speeds of 150,000 kilometers per second, which is a little over half the speed of light.

This is a rather generous speed, considering that the fastest craft ever developed by humans is Voyager 1, which is currently traveling at about 17 kilometers per second, or 0.000058% the speed of light.[125] A speed of 150,000 km/s is approximately 8,800 times faster than Voyager 1 and 420,000 times faster than a commercial airliner. However, even at this blistering speed of about 93,150 miles per second, those who stay behind on Earth would still have to wait nearly 100,000 years for the astronauts to return from their mission to the M13 star cluster. Additionally, following the 50,000-year return trip to Earth, the distant descendants of the astronauts who had originally met with the M13 civilization would only be able to describe a very ancient form of those beings, if they still exist at all.

A 100,000-year roundtrip space voyage does not seem feasible, particularly considering that this is about 2,000 times longer than the average human lifespan and about half the entire history of our species, as we developed into anatomically modern *Homo sapiens sapiens* around 200,000 years ago. In order to complete such a journey, we would have to find some way of slowing senescence, the process of biological aging, or we would just have to keep reproducing onboard the spacecraft and rely on subsequent generations to do the meet and greet with our celestial neighbors. However, as the overabundance of outlandish reality television shows continually demonstrates, anytime human sexual behavior is added to a situation in which people are living in tight quarters, things tend to get complicated and often hostile. The need for in-flight sexual reproduction may be an inevitability of long-distance space travel, given that our arrival on a distant planet may require that we keep making baby astronauts to replace the ones that came before, regardless of the social consequences.

This hypothetical voyage to M13 would naturally be much shorter if we were able to go faster. For instance, at sustained speeds closer to 285,000 km/s, or about 95% the speed of light, a round-trip visit to M13 would "only" take about 52,500 years to those on Earth and about 16,283 years to the astronauts onboard the spacecraft, due to the effects of time dilation. Although this would make the excursion a substantial number of bathroom breaks shorter, it may never be possible to travel at such speeds. In fact, even half the speed of light is a generous estimate for how fast we might be able to travel through space in even the next 5,000 years.

In addition to showing how space and time become warped for

individuals traveling at high speed relative to each other and to the speed of light, Einstein's theory of relativity specifies that it is impossible for anything with mass to match or beat this speed. This is because the mass of an object becomes infinite as it approaches the 299,792.46 kilometer-per-second speed limit of the universe, meaning that inertia acts against objects with increasing force as they move closer to the speed of light. Light can move this fast because it is massless as a wave, but for all other bodies that by definition have some quantifiable mass (and particularly a spacecraft with crew, equipment, and supplies for a 16,000-year round-trip voyage), the inertial forces working against it, even in the vacuum of space, may be impossible to overcome at even a fraction of the speed of light.

Although it is popular in science fiction films, the concept of a *warp drive*—a hypothetical faster-than-light propulsion system to enable interstellar travel—has been deemed unrealistic by even the most optimistic experts.[126] For example, Physicist Marc Millis, who led the Breakthrough Propulsion Physics Project at the Glenn Research Center for NASA, considers it somewhat of an "idea zombie." In a recent interview for *Discover Magazine*, he is quoted as saying:

> The EM-Drive and warp-drive stuff is not in the category of what I'd call 'behaving scientifically,' Millis sighs when I mention the Drive. He has a whole category for "idea zombies—bad ideas that won't stay in their graves." In an influential review paper, he picked through more than a dozen proposed interstellar propulsion technologies and ruled out many of them as either improbable or impossible.[127]

This hypothetical space-travel scenario centered on the Messier 13 globular cluster in the Hercules constellation, which is about 25,000 light-years, or about 147,000 trillion miles from Earth. This is seemingly very far away, although in comparison to the rest of the universe, it is actually quite close. To put it in perspective, our Milky Way galaxy is upwards of 100,000 light-years across, and there are tens of billions of other galaxies billions of light-years beyond it. Traveling to those more distant stars and galaxies would naturally present even greater challenges, though even attempting to visit one of the recently discovered Kepler-62 planets at 1,200 light-years from Earth, or Kepler-186f at 500 light-years, or even Kepler-438b at 475 light-years (2,792,347,051,195,674 miles), would present all of these same challenges, just on a slightly smaller scale.

Interstellar travel to any distant or nearby earth-like planet is not likely to happen anytime soon. This is not because we lack the ingenuity to develop fast spacecraft, but simply because those craft are bound by the laws of physics, meaning that travel to and from them may always require a tremendous amount of fuel, equipment, supplies, and perhaps most importantly, time. In fact, Marc Millis, the above-mentioned physicist researching the logistics of interstellar travel, states in the same Discover Magazine interview that 71 years might be the upper limit for any interstellar astronaut.[127]

Even if it were logistically feasible to travel in a small craft at high speed for 71 years, upon returning to Earth, everyone the astronaut knew would be long-dead, and they would be very out of touch with their own civilization as a result of the future time travel effects of time dilation. Because of the universality of the laws of physics, it is safe to assume that other lifeforms on other planets who may wish to venture out beyond their solar vicinity would also experience these same problems associated with long-distance spaceflight.

Realistically, it may be impossible to find anyone who would be willing to sacrifice their entire lives, as well as those of their children, and perhaps grandchildren depending on the length of the journey, in order to undertake such a mission. There are certainly many logistical problems associated with long-distance space travel, but we must also consider the emotional and psychological hardships, as well as the personal, familial, and social sacrifices that would need to be made in order to carry out any interstellar mission.

Compared to interstellar travel, a backward time travel mission would not require the same time commitment. For instance, imagine a group of future anthropologists, medical examiners, linguists, etc., preparing to head back in time to collect biological and ethnographic data on an ancient civilization living in the year 2,418 AD. They kiss their husbands, wives, and children goodbye, and then head off to the time dock to board their rotating cylinder at 9:00 a.m. Even if this mission was scheduled to last two weeks, because of the nature of time and time travel, there is no reason why they couldn't return to their home time at 5:00 p.m. on the day they left, or at 9:05 a.m. for that matter. If the researchers were to do this often enough they would eventually begin to look older at a faster rate than those who continued their more linear path through time. However, this would not be a multi-decade or multi-century mission, as would be the case with interstellar researchers, who may be required to commit the remainder of their lives to an investigation of deep space.

With regard to the question of interstellar travel, it is important to consider the physical laws of the universe, the immense distance between stars and planets, the incredible amount of time it would take to travel to them, the logistical difficulties of transporting people and cargo across the cosmos, as well as the psychological and social hardships associated with such a journey. Because of these and other physical, emotional, and cosmic constraints, it is hard to fathom that extraterrestrials have or ever would be able to visit Earth from a distant planet in a distant star system elsewhere in the universe.

5.2. General Relativity, Warped Spacetime, Hyperspace, and Wormholes

When describing to people all of the inherent difficulties associated with interstellar travel and how unlikely it is that the UFO phenomenon can be explained by extraterrestrial visitors, they often bring up the theoretical notion of warp drives and wormholes as a way to shorten the commute. These hypothetical methods of deep-space travel still do not address the important question of how two civilizations would ever find each other in the vast emptiness of space. Though even beyond this and other related issues, there are numerous physical and logistical problems associated with these proposed mechanisms for reducing the time commitment necessary for interstellar travel.

As mentioned above, experts do not consider warp drives to be a realistic means of deep-space exploration; so what of wormholes? Also known as an Einstein-Rosen Bridge, wormholes are a hypothetical aperture that acts as a shortcut between two separate regions of spacetime. Wormholes are theoretically possible, as demonstrated by Einstein and Rosen in the 1935 paper *The Particle Problem in the General Theory of Relativity*.[128] However, a more important issue relates to whether or not an individual and their spacecraft would ever be able to survive passing through one, due to tremendous inertial and gravitational forces in and around it.

Einstein's theory of general relativity, which followed ten years after his theory of special relativity, introduced the notion of curved spacetime. He demonstrated how massive bodies can bend and twist the fabric of space and time, influencing the movement of neighboring objects, including light. For instance, light radiating out from a distant object will be bent by the gravitational field of other massive bodies (e.g., distant galaxies, planets, and dark matter), as it travels toward our location here on Earth, in what is known as gravitational lensing. In fact, in addition to the transit, wobble, and direct imaging methods, the light-bending effect of distant celestial bodies is used by astronomers to look for new exoplanets, in what is referred to as the microlensing method.[129]

Because space and time are thought to be fused into a four-dimensional continuum, the warping of space by massive bodies also elicits an effect on time. In the above example of long-distance space travel to M13, it is apparent how time dilation occurs in the inertial reference frame of someone traveling at high speed relative to the speed of light and to those who remained on Earth. This same time dilation effect would occur if an individual simply visited a very large/dense planet or hung out near a black hole for a while. In fact, if you were to return to Earth after living on the moon for a few years, you would find that all of your clocks are slightly faster than those of your terrestrial counterparts. This is simply because there is less gravity on the moon, which dilates time and slows clocks to a lesser extent than on Earth. In essence, space and time are relative to how fast you are moving and how much gravity you are experiencing at any given time.

More massive objects like black holes elicit a much greater effect on surrounding matter in the universe, warping it to such an extent that it may be possible to connect different regions of spacetime, if one were able to manipulate this warping effect. Dr. Kip S. Thorne, Feynman Professor of Theoretical Physics at the California Institute of Technology, is a leading researcher on black holes and wormholes. In 1985, he was asked by Carl Sagan to review the manuscript for his forthcoming novel *Contact*, in the hopes that Thorne would be able to fix any obvious contradictions to the laws of physics in order to make the literary first meeting between humans and extraterrestrials as realistic as possible.[130] In Sagan's original manuscript, the Earth astronaut reached this faraway civilization by means of a black hole that connected their two distant regions of space. However, as Thorne pointed out to Sagan, if someone were to attempt to pass through the singularity of a black hole (i.e., the point at which matter becomes infinitely dense within it) they would be compressed and stretched by tidal forces, which would rip them into increasingly smaller pieces:

> It is utterly impossible for the astronaut to move on through the singularity and come out the other side . . . all astronauts, particles, waves, whatever, that hit it are instantaneously destroyed, according to Einstein's general relativistic laws. . . . Not only is an astronaut stretched and squeezed infinitely at the singularity, according to the Oppenheimer-Snyder equations; all forms of matter are infinitely

stretched and squeezed—even an individual atom; even the elec-
trons, protons, and neutrons that make up atoms; even the quarks
that make up protons and neutrons.[131]

Because of the quark-shredding properties of a black hole, Thorne
suggested that Sagan instead use a wormhole as a way of getting the
astronaut from Earth to this distant civilization. However, while death
by a thousand cleaves does not seem as inevitable in a wormhole—at
least compared to diving into the event horizon of a black hole—there is
still a real danger of being torn apart by one. Warping spacetime to the
extent that it connects two disparate regions of space would require an
enormous amount of mass and energy, which could potentially rip apart
a spacecraft and its occupants. Additionally, mathematical calculations
derived from Einstein's field equations suggest that wormholes are short-
lived and highly unstable, and amplification of electromagnetic radiation
at either mouth of the wormhole could destroy it shortly after inception,
making any attempt at passing through it very dangerous.[132]

In his 1995 book *Black Holes & Time Warps: Einstein's Outrageous Legacy*,
Thorne describes how he mulled over the problem of creating and maintaining a wormhole so that some advanced civilization could use it for
interplanetary travel, without it collapsing or killing them by some other
means.[132] In order for this to be achieved, Thorne postulated that it would
first be necessary to connect two points in *hyperspace*, which is a theoretical
concept advocating that our three dimensions of space are embedded
in six dimensions of hyperspace. In theory, this hypothetical hyperspace
connection could be carried out using two supermassive bodies capable
of bending these separate regions of spacetime toward each other.[133] The
two supermassive bodies would then need to be connected by a tunnel
and, importantly, all of these components would need to be made stable
in order to avoid its collapse. To help keep the wormhole open, Thorne
and his graduate students proposed that it be filled with a kind of material
that would push the walls apart gravitationally. This *exotic material* as they
called it, would produce a sort of anti-gravitation effect, using the negative
pressure of a vacuum to produce a gravitational repulsion force, similar to
that which began pushing all of the matter in the universe apart during
the Big Bang, around 14 billion years ago.[132]

The principle limitations regarding the theoretical use of wormholes
to bridge great swathes of the universe center on how a civilization could
move or manipulate supermassive bodies, how long it would take for them
to travel to these massive objects, how they could create a tunnel to con-

nect them, and how they could pass through a dense and super-energetic tunnel through hypothetical hyperspace without being crushed or ripped apart. As per Newton's third law, the force exerted by the exotic material used to prop open the tunnel must be at least equal to, and opposite, the force of gravity of the singularity-like mass used to form a bridge through spacetime. So attempting to pass through matter with these properties could potentially be like trying to shove a sperm whale through a ball bearing, or a tomato through a screen door.

Even if a wormhole were to produce an actual hole through spacetime, such as that which is thought to exist at the event horizon of a black hole, we may not be able to accurately predict what happens after someone passes through it, given that they would suddenly exist in a separate place in the universe...and at a different time. It is possible that they could be torn apart inside, while still appearing to hover near the mouth of the hole, such as that which is predicted to occur at the event horizon of a black hole, where light becomes trapped as it approaches the point at which the escape velocity of the black hole reaches the speed of light. They could also potentially exit the wormhole at the beginning of spacetime itself, destined to be reborn with all the matter in the universe at its cyclic origins during the next Big Bang.

In the best-case scenario, if the astronaut did happen to make it through the wormhole with atoms intact, it would be difficult to predict where and when they would end up on the other side and whether they would ever be able to return to their original position in spacetime. Accurately and reliably sending someone across large expanses of space and through long periods of time would require a high level of control over the specific physical parameters of a wormhole-creating device. This must surely be a difficult beast to tame to boot, as this may require the manipulation of masses and energies somewhat akin to that which exists at the center of our galaxy.

At the present time, what is needed to construct a functional wormhole does not exist in the natural universe, which may mean that the idea remains theoretical for some time to come. Wormholes are not necessarily "idea zombies" as in the case with warp drive technology, though from a functional space and time travel standpoint, they face a number of important challenges. In fact, because of current limitations in our understanding of wormholes, Thorne commonly refers to those who may one day construct one as being part of an *infinitely advanced civilization*, which he defines as "one whose activities are limited only by the laws of physics, and not at all by ineptness, lack of know-how, or anything else."[134] In other words, if the laws governing the universe do not absolutely prohibit it, then it should be considered possible, and someday an extremely advanced civilization may figure out a way to do it. However, given the tremendous forces involved and the necessity for keen and precise manipulation of them, we may be waiting for some time. Additionally, because all civilizations may only have 1,000 to 100,000,000 years to reach such a state of high intellect before destroying themselves—as per "L" in the Drake equation—the clock is ticking for us, as well as any other lifeform that may be on course to becoming infinitely advanced.

5.3. The Complexity of Interstellar Communication

5.3.1. Lost in Space

Unless we get a response from one of the many radio signals we have been broadcasting into space since 1974 as part of the ongoing Messaging to Extraterrestrial Intelligence (METI) and Search for Extraterrestrial Intelligence (SETI) programs, there is little chance of actually finding intelligent life on other planets. The vast distance that separates solar systems prohibits in-person planet-by-planet searches, and beyond scanning exoplanets for biosignatures, we are mostly left with sending and waiting to receive messages from what, if any, complex lifeforms might exist beyond Earth. However, even this far-simpler method of seeking out extraterrestrial intelligent life is riddled with hitches. For instance, any signal broadcast into space faces the same challenge of having to cut through the omnipresent background radiation and celestial noise that permeates the universe. Unless an intelligent alien civilization began broadcasting long ago with a very strong signal at a frequency we can detect, and which is able to cut through all of this background radiation, it may be very difficult to distinguish the signal from all of the other noise in space, even on the off-chance that it was aimed directly at us.

We are able to minimize many of the frequencies that create this background noise, so as to better hear a signal that might be coming through. However, these interfering frequencies might not be the same everywhere in the universe. In other words, if another civilization also learned how and why to do this, the signal could still be lost if it arrives with the wrong frequency and amplitude for the specific background radiation of their solar locality. Additionally, depending on the type of matter that lies between us and them—be it a black hole, pulsar, dark matter, etc.—the signal could become heavily distorted or potentially destroyed altogether.

The amount of money spent by METI and SETI in their search for intelligent extraterrestrial life seems exorbitant, considering that all this investment is likely to return is false hope. While out fishing, it is easy to convince yourself that just one last cast could be the one that lands a monster fish, and occasionally it does, although fishing in water on Earth has been shown to deliver far more tangible results than scanning the skies for something that may not exist at all. Searching the heavens above for another lonely lifeform in the nearby galaxy may be important in that it makes us feel as though we are trying, while providing a sense that we may someday be less galactically lonely. However, given the improbability of complex life arising at the same time and in nearly the same place among the 100 billion stars in the universe, it can also feel as though we are squandering resources on an ephemeral search for something that may not be there to be found.

5.3.2. *Lost in Translation*

There are many reasons why our search for intelligent life on other planets is not likely to return anything meaningful anytime soon, or ever. Though if by some miracle we did happen to locate another intelligent civilization living around a nearby star, we may face a whole new set of challenges. Once again, there is the all-important issue of time and the amount of it that we would waste sitting and waiting for transmissions to be sent and received following initial contact. Not to mention how frustrating it would be trying to learn each other's symbolic language with multi-year, decade, or century-long breaks between transmissions.

This is a common problem even while communicating with fellow humans on our own planet. Broadcasters must cope with long pauses and constant unintended interruptions, due to the added time it takes to send and receive signals on opposite ends of the Earth. This communication delay is even more apparent in interviews with astronauts aboard the International Space Station. Watching live-television interviews with those aboard can be quite painful due to the delay between questions and answers, particularly considering our recently acquired species-wide attention deficit disorder, which itself is partly tied to technological advances over the last 30 years that have helped accelerate the speed of communication. In a chapter of *Archaeology, Anthropology and Interstellar Communication*, titled *Reconstructing Distant Civilizations and Encountering Alien Cultures*, METI president Douglas Vakoch writes:

> Even if we detect a civilization circling one of our nearest stellar
> neighbors, its signals will have traversed trillions of miles, reaching
> Earth after traveling for years. Using a more sober estimate of the
> prevalence of life in the universe, our closest interstellar interlocutors
> may be so remote from Earth that their signals would take centuries
> or millennia to reach us.[135]

Another problem we are likely to face involves the challenge of communicating with beings that evolved on a separate planet, who may not
possess the same functional anatomy for hearing, seeing, or vocalization.
As mentioned previously, the physiology of extraterrestrial organisms
would undoubtedly be fundamentally different from our own, simply due
to different environments eliciting different selective pressures throughout
their evolutionary past. However, even if extraterrestrials do not evolve
anything akin to our human ears, larynxes, brains, and mouths, their
form of communication is still likely to involve vibrational sensitivity, as
is common here on Earth. In fact, Seth Horowitz, a neuroscientist and
author of the book *The Universal Sense: How Hearing Shapes the Mind*,[136]
believes vibration-based communication will develop everywhere that
life arises in the universe:

> Vibration sensitivity is found in even the most primitive life forms,
> even bacteria. It's so critical to your environment, knowing that some-
> thing else is moving near you, whether it's a predator or it's food. ...
> If we find life on other planets—if it's more complex than microbes
> or viruses—they'll have vibrational sensitivity.[137]

Vibrational sensitivity is certainly widespread on Earth, although—even
if it is also common throughout the universe—this does not necessarily
mean that terrestrial humans and extraterrestrial non-human lifeforms
would possess the same functional features to send and receive mutually
understandable messages. For instance, what if the extraterrestrials buzzed
and danced like bees, used infrasound like an elephant, ultrasound like a
tarsier, seismic communication like the demon African mole rat, or any of
the various other forms of vibrational communication observable among
species here on Earth? We should be able to hear, feel, or use technology to detect communicatory vibrations made by extraterrestrial aliens.
However, establishing symbolic meaning from these vibrations in order
to understand their language and the meaning behind what is being said
would be a far more difficult challenge.

There are between 6,000–7,000 languages spoken around the world
today. Each is easily recognizable as human language and, with some

effort, they can all be learned. This is simply because they are spoken by the same species, which has evolved the same respiratory, vocal, lingual, auditory, facial, and cerebral anatomy that facilitates this highly complex form of linguistic speech and understanding. We are sensitive to, or can at least detect with auditory devices, the vibrational communications of most other animals. However, we remain unable to translate and understand what it is they are actually communicating, despite using it ourselves and despite having lived alongside these animals for millions of years on the same planet. In fact, it took until the 1960s before we realized that humans could engage in any form of meaningful communication with other animals here on Earth. In 1967, Beatrix and Allen Gardner began a language acquisition study at the University of Nevada, Reno, which aimed to teach American Sign Language to a chimpanzee named Washoe.[138] After some effort, the researchers, along with most of the world, were amazed to discover that Washoe was able to learn 350 signs and even taught her adopted son some of these signs, independent of the researchers.

It should be noted that although humans share more physical and cognitive traits with chimpanzees than any other animal on this planet, we are still unable to speak directly with them. This is not because chimpanzees lack the mental capacity for complex speech, rather, it is because their vocal tract anatomy is different from our own, which makes them physically unable to produce the same sounds that we use as part of our oral language. This difference in linguistic abilities is largely a product of our bipedalism, which modified hominin vocal tract anatomy in concert with other craniofacial changes as we moved toward habitual upright walking.

As described previously, standing upright meant that our heads needed to rotate downward in order to see where we were going. This reconfigured our mid and lower faces, palates, airways, and vocal chords, pushing our larynx steadily downward over the last 6 million years. Our more inferiorly positioned larynx granted us the exceptional ability to vocalize a broader range of sounds and, in association with our larger and more complex brains, helped facilitate what has become the most adept form of communication on this planet.

Humans and chimps are two recently divergent and remarkably similar species, yet it still took 99.9999996% of the history of life on Earth before we, or any other advanced species, became capable of complex communication with each other. Furthermore, to date, this has really been a one-sided affair, in which we taught chimps to speak to us in our own language, but we remain ignorant of what chimps are saying to each other, or to us when not using the language we taught them. Essentially all we did was force them to learn our symbolic language in a rather anthropo-centric and narcissistic manner, while making no effort to learn how to speak or understand their native hominoid language. Though if we were to ever make contact with intelligent extraterrestrial beings, this colonial mindset and system of forced assimilation may need to be replaced by one that is a bit more collaborative and mutually lucid.

Assuming an alien civilization happened to evolve something even re-motely similar to what we use for vocalized vibration-based communica-tion, it would take an inordinate amount of time in the beginning stages of contact to teach and learn each other's languages. This learning curve becomes exponentially steeper if our initial linguistic instructions are only able to come via interstellar radio signals, considering there would be decade or century-long delays between transmissions. It could potentially take hundreds or thousands of years to learn even rudimentary elements of an extraterrestrial's language, depending on where their planet may be located in the galaxy, and how much, if anything, could be garnered from each transmission. According to anthropologist Ben Finney, historian Jerry Bentley, and METI president Douglas Vakoch:

> To state what may be obvious, if we receive a message from ex-traterrestrials, we cannot count on their providing direct translations from one of their native languages to any terrestrial language. ... We may be able to understand basic mathematics and astronomy, but once extraterrestrials begin to describe their cultures, interstellar comprehension may suffer considerably.[139]

It is even difficult to decipher the language of past human groups on this planet, including those that lived as recently as 3,000 years ago. Without any context or shared words or phrases between modern and ancient lan-guages, it is almost impossible to know what a culture's arbitrarily chosen symbols were meant to represent. In fact, without the Rosetta stone, we would still only have a vague idea of what was being communicated in Egyptian hieroglyphics.

Discovered by Napoleon's army during an Egyptian military campaign in 1799, the Rosetta stone was critical to deciphering Egyptian hiero-glyphics (figure 5-2). This is because, rather fortunately, the Rosetta stone happened to contain the same text, written three different times, in three different languages. Because 19th century European linguists were able to read and understand one of these languages, they could then work out what was being said in the other two, and eventually decipher the entire writing system for these previously unknown languages.[140]

Figure 5-2. *The Rosetta stone*

Considering all of the difficulties in translating the languages of recent human civilizations, despite their being spoken by members of the same species, on the same planet, with the same brains and functional anatomy for speech and hearing, it would undoubtedly be much harder to decipher any communication sent to us by an extraterrestrial civilization. Without context, associated meaning, or anything universally relatable, we aren't likely to understand an alien message, and they aren't likely to understand ours either.

With this limitation in mind, our first major attempts at communicating with extraterrestrials kept the message relatively simple and primarily included images and basic aspects of science and mathematics that were critical to translating ancient Mayan inscriptions.[140] When the Pioneer 10 and 11 spacecrafts were launched in 1972 and 1973, respectively, they carried with them a gold plaque. This ornamental tablet described what humans look like, where we are in the galaxy, the date the mission began, as well as an illustration of the common element hydrogen, which was thought to be easily translatable by anyone from a scientifically literate civilization.[141]

When Voyager 1 and 2 were launched in 1977, they carried a 12-inch gold-plated phonograph record that contained more information in various formats. For instance, the record included 115 images and spoken greetings from a variety of different languages, beginning with a 6,000-year-old Sumerian language called Akkadian and ending with the modern Chinese dialect Wu. It also included 90 minutes of an eclectic mix of music, as well as a variety of nature sounds representative of Earth.[142] These images and sounds, assembled by Dr. Carl Sagan and a number of his associates—including his second wife Linda Salzman Sagan—were meant to convey something about humans and our planet in the very remote chance that they would be discovered by extraterrestrial intelligent beings.

Sending radio signals and unmanned probes into outer space seems rather pointless if there are no civilizations out there to receive them. Though even if there are advanced lifeforms nearby who are capable of receiving, understanding, and responding to an interstellar message, we still face an uphill battle. Numerous logistical difficulties are inherent in attempting to communicate with beings on a different planet, far from Earth, who may not have evolved anything analogous to our eyes, ears, mouths and brains. In fact, we are likely to share far fewer morphological and cognitive characteristics with extraterrestrials than the mammals on our own planet, which we are still unable to communicate with in any meaningful way.

5.4. The Relative Ease of Intertemporal Exchange

Turning to the question of IFOs and extratempestrials, one can appreciate how future humans would have a much easier time communicating with us in their own past, as a result of cultural continuity and our shared history on the same planet. Unlike extraterrestrial aliens, our descendants would benefit from being able to look back through historic records and other forms of preservable media to aid them in learning the same languages that were once spoken by their own ancestors. The relative ease of intertemporal vs. interplanetary dialog may also be found in the effortlessness with which communication takes place—in the individual's native tongue—in reports of abductions and other instances of close encounters.

For instance, in the accounts of Betty and Barney Hill, both describe being spoken to in their native language of English. More specifically, when Barney was on the road awaiting abduction, he was told: "keep looking and that no harm would be done to you." Betty also recalled that the "examiner" who performed her medical examination spoke in broken English compared to the others aboard, who seemed to have a better mastery of the language.[143] This is just one example among many different reports that describe extratempestrials speaking in the regional language of those they encounter in our time.

Interestingly, this cross-temporal communication takes place even though it is highly unlikely that extratempestrials themselves will still speak any of our current languages in the distant future. This is indicated by the rapid rate of language evolution, as well as the ubiquity of close encounters describing "strange symbols and writing" in and on IFOs. Though even if no contemporary languages are still spoken on Earth in the distant future when we begin to travel through time, our descendants would at least be afforded the luxury of being able to look into their own past for guidance on how to relearn a dead, or highly divergent earlier language before visiting us in the past.

Becoming proficient in a group's language helps facilitate data collection and aids in reducing bias that can result from miscommunication and misunderstanding. In fact, during the early 1900s, Franz Boaz, who is considered the father of American anthropology, began to require that his students learn the language of the specific cultural group they intended to study. Margaret Mead was one of his first students to approach ethnographic research in this way. Her work on the island of Samoa in the mid-1920s, and her subsequent publication of the book *Coming of Age in Samoa*, proved to be instrumental in redefining the way we view the roles of culture and biology in shaping human behavior during adolescence.[144]

Without a finely tuned mastery of the local language, this ethnographic study could not have generated the same insights, since we cannot fully appreciate the intricacies of a cultural group without a complete understanding of their mode of communication. Betty and Barney Hill's abduction account, along with hundreds of other vetted reports of close encounters involving direct communication in the abductee's mother tongue, suggest that we will continue Boaz's best practices for learning the language of the specific cultural group—and of the specific time period—prior to conducting ethnographic field research in the future.

These IFO reports also indicate that some occupational, or role-specific stratification, may exist among the members of these interdimensional research expeditions. For instance, Betty Hill stated that the "examiner" did not possess a mastery of the English language, which made him harder to understand compared to other individuals she interacted with while onboard the IFO. This linguistic disparity suggests that the examiner may have occupied a more specific role tied to conducting physical examinations and that his English ineptness could be mitigated by the linguistically literate chaperones aboard, whose more interactive roles would require that they possess some expertise in the language and culture of mid-20th century America.

Supposing that at least some of the individuals aboard an IFO are anthropologists—which seems likely considering that investigating humans in the past and present is exactly what we do—then they too may specialize in specific hominin species, cultures, languages, and historic and prehistoric periods of the past as we do now. With this in mind, extratempestrial anthropologists encountered in our present are likely to be experts on the physical and cultural characteristics of humans who exist at this specific place and time in their past. The main difference between us and them, is that modern anthropologists are only capable of conducting research into the material culture and morphological form of long-dead civilizations, using anthropometrics, genetics, and archaeological methods that are far more primitive by comparison. However, our cross-temporal descendants would be afforded the luxury of working as biological, cultural, and linguistic anthropologists, conducting morphological and ethnographic research on past peoples while they are still alive.

If we are someday able to harness the data-generating power of backward time travel technology, our knowledge and means of investigating hominin history and prehistory will undoubtedly advance far beyond that which exists today. Biological anthropologists, and particularly paleoanthropologists like myself, would benefit enormously from the analysis of soft-tissue anatomy, considering that we are currently only able to examine the teeth and fossilized skeletal remains of our distant hominin ancestors. Cultural anthropologists would also be major beneficiaries of a time travel device, as they would suddenly be granted the opportunity to conduct real-time analyses of an entire culture, taking holistic ethnographic methods used for researching living peoples and extending them deep into the human past.

5.5. Anthropological Spacetime

It is estimated that there are between 100 billion and 400 billion stars in the Milky Way galaxy,[145] and perhaps as many as 100 billion to 200 billion galaxies in the universe.[146] There is also an incredible amount of space, and therefore time, separating all of the galaxies and solar systems throughout the universe. Because of the colossal amount of time that it would take to send and receive interstellar messages, and the even longer time necessary to actually visit distant star systems, intelligent life could evolve on a planet, become complex enough to send radio signals out into space, and then completely die out before any of their signals reached another interstellar civilization.

I am not at all skeptical about the possibility of life, or even complex life, evolving on another planet. Although, it is somewhat incomprehensible how we would ever be able to find each other in this massive and continually expanding universe. I am also unconvinced that an extraterrestrial intelligent lifeform could ever evolve to look anything like us, or that we would have the capacity to understand each other's form of communication. Additionally, even if we were to make contact, the amount of time wasted sending and waiting on transmissions and trying to figure out what each other is saying may prohibit any meaningful exchange from being realized at all.

Because space is so immense, and there are so many planets in the habitable zone of their star, it is difficult to discount the possibility that life is evolving on some of them right now. After all, it already happened once here on Earth, soon after this planet became capable of sustaining life. This indicates that it can, and most likely will, happen again somewhere else. However, considering all of the challenges associated with interstellar space travel and communication, perhaps it is time we move past the idea that close encounters with alien beings—which look, act, and speak like us—have anything at all to do with other lifeforms on other planets.

There are numerous problems associated with locating, communicating with, and traversing huge distances in order to visit other civilizations. There are also too many common features between humans and the beings described in IFO reports for any consideration to be given to the idea that we are being visited by creatures from a distant planet. In fact, this may represent the best line of evidence against the long-standing notion that IFOs and aliens are the product of a separate biocultural evolutionary process on a different planet. For if they are indeed real, the overwhelming similarity between us and the reported them could only be attributed to a deep-rooted, shared evolutionary history, here on planet Earth.

6

For What Is Time?

For what is time? Who is able so much as in thought to comprehend it, so as to express himself concerning it? And yet what in our usual discourse do we more familiarly and knowingly make mention of than time?[147]

– Saint Augustine

6.1. Limits of Our Current Understanding of Time and Time Travel

Time, in both the physical and philosophical sense, remains a poorly understood aspect of our universe. In any field, when something pivotal is not yet fully appreciated by those with a stake in unraveling the mysteries that surround it, differing views and divergent ideological camps often develop. This is certainly the case regarding the physics and metaphysics of time and to an even greater extent with respect to the specific question of backward time travel. Notably, differences of opinion abound concerning whether or not humans will someday develop the knowledge and machinery necessary to achieve backward time travel, despite broad consensus that time travel to the future is simple.

The specific question of time travel to the past has been approached with cautious optimism by many researchers, though occasional staunch unwavering pessimism by others. However, some common ground has begun to emerge between both camps. This area of agreement largely centers on the notion that we cannot fully understand if backward time travel is possible until we are able to merge Einstein's Classical General Theory of Relativity (GTR) with quantum mechanics. This is because some solutions to Einstein's field equations allow for *Closed Timelike Curves* (CTCs), which are formed through the bending and warping of spacetime in such a way that travel to the past becomes possible. However, things get a bit messy with the formation of CTCs using GTR models, particularly with regard to paradoxes that seem to mar cause and effect. However, in quantum mechanics, these cause/effect paradoxes are less problematic, as they consistently reveal an inherent self-consistency among events, which are, and have always been, intricately linked across different regions of spacetime.[148] [149] [150]

The notion of (and mathematical evidence for) self-consistency among different events connected by CTCs, was originally introduced by renowned Russian physicist Igor Dmitriyevich Novikov.[151] As such, this later become known as the Novikov Self-Consistency Principle:

> For spacetimes with CTCs, past and future are no longer 'globally' distinct…events on CTCs should causally influence each other along the 'loops in time' in a self-adjusted, consistent way. This requirement has been explicitly formulated as the 'Principle of self-consistency,' according to which the only solutions to the laws of physics that can occur locally in the real universe are those which are globally self-consistent.[152]

Until general relativity and quantum mechanics are united in one grand theory of everything or replaced with something more holistic that is not yet known, contentious rows regarding the question of time travel to the past are likely to persist. Though in the meantime, some insight may be realized from taking an alternative approach to this question, by instead asking if there is anything in the laws of the universe, based on our current knowledge, that would prohibit time travel to the past.

A paper by Earman et al. (2009) titled *Do the Laws of Physics Forbid the Operation of Time Machines?* sought an answer to this question of whether backward time travel is *im*possible, using a broad approach rooted in physics, history, and philosophy. This review paper examined the vast amount of literature concerning backward time travel research, with a specific focus on more critical studies whose results suggest that time travel to the past is not possible. This included more skeptical approaches like Stephen Hawking's Chronology Protection conjecture (which is the idea that the laws of physics prevent time travel for macroscopic objects as a result of perceived causality violations),[153] as well as similar other disparaging studies of time machines under classical general relativity, semi-classical quantum gravity, quantum field theory on curved spacetime, and Euclidean quantum gravity.[154] The results of this research, as well as an earlier investigation by Earman (1995),[155] demonstrated that these anti-time travel studies, including Hawking's Chronology Protection Conjecture, are not actually as prohibitive as they may seem:

> Our verdict on the question of our title is that no result of sufficient generality to underwrite a confident 'yes' has been proven.[154]

Although, these authors also highlight the current difficulty associated with assessing the potentiality of backward time travel, devoid of a unified theory of gravity and quantum mechanics:

> Perhaps in the not too distant future it will be possible to discuss the implications of a full quantum theory of gravity for time travel and time machines. But that day is not here.[154]

This review article suggests that nothing in nature forbids time travel to the past for macroscopic objects. However, as the authors of this and similar other studies commonly point out, numerous limits to our available knowledge exist and, perhaps somewhat ironically, largely as a result of our less lucid position in time. Despite these shared limitations, the above paper is important, in that it brings to light a perceived disconnect between studies of the physics of time travel, and the logical and philosophical issues that arise if someone were to actually visit a past period of time:

> ...the physics literature on time machines engages a set of issues that are largely distinct from those involved in the so-called paradoxes of time travel. More attention to the former by philosophers could reinvigorate discussions of the paradoxes and related issues. ...The study of time machines is a good opportunity for forging a partnership between physics and the philosophy of physics.[154]

It is important to consider the whole picture in assessing an endeavor as complex as time travel. As such, researchers in the disciplines of physics, mathematics, astronomy, philosophy, and certainly anthropology, cannot work in isolation. In fact, an awareness of the broader scope of humanity, as well as one that considers instances of intertemporal interaction that may already be occurring among humans from different times, could advance our understanding of the physics and engineering necessary to construct a backward time travel device, as well as what may occur and what may be learned from using it.

The need for maintaining a strong working relationship among disciplines is becoming increasingly important as our knowledge of time and space deepens. Ideological conflicts between fields, between disciplines within the same field, and among individuals within and among different branches of science, are often roadblocks to the furtherance of knowledge. However, all perspectives and approaches should be welcomed and encouraged as we pursue a deeper understanding of time and the mechanisms that may someday allow us to move though it in new and insightful ways.

Although there is still much we do not understand regarding the structure of time and space, most scholars are likely to agree that human ingenuity is unbounded. Looking back through our long hominin history, it is clear that we have come a long way in developing an advanced intellect and unique ability to modify our environment to accommodate whatever need or desire we possess. Human innovations have continued to accelerate over recent periods of the past as well, which suggests that this advancing rate of intellectual and technological development will continue into the future. With it—as with our ability to understand the mysteries of the moon, comets, exoplanets, and distant space beyond our solar system—so too may we begin to resolve the perplexity of time, so as to someday realize the grand achievement of returning to the past.

6.1.1. Bridging the Hominin Past, Present, and Future

The concept of time—and our sense of its seemingly relentless unilineal passage from past to future—has always been known to us in some capacity. But it is only in the last 100 years that we have begun to unravel the mysteries of this entity that is so integral to our everyday lives. This is largely the result of Einstein's contribution of special and general relativity, as well as subsequent research in quantum mechanics, string theory, loop quantum gravity, and others, which have helped to broaden our understanding of time.

Ongoing advancements in our knowledge of time, and indications that nothing in nature prohibits people and things from roving amongst different periods of it, provides important context for understanding how our more advanced progeny may someday achieve this monumental feat. More specifically, deeper appreciation of the physics of time and time travel aids in elucidating why the big-headed, big-eyed, small-faced, tailless, bipedal beings reportedly encountered by modern and past humans throughout history and prehistory should be recognized as our distant human descendants.

An ever-broadening comprehension of space and time—in the context of both physics and long-term hominin biocultural trends—indicates that time travel may be a better explanation for this phenomenon. If this is someday proven to be true, given that it is a testable and falsifiable theory, the stark morphological differences between us at the present time and them at a future time suggests that this eventuality exists far to our future. As such, any description of our current understanding of time, and ongoing research into the physics of backward time travel, is likely to seem extremely primitive and uninformed from the perspective of our distant descendants, who by that time will have long-since unraveled the enigmas we ponder today.

Nevertheless, given the multiplicative nature of technological change, whatever knowledge we possess now will likely still be relevant in the future, as culture and biology are always built upon a preceding and often less complex form. Subsequently, and cautious of these temporal limitations, the narrative turns to providing a snapshot in time of our current understanding of time. This deeper delve into contemporary temporal comprehension provides a sense of how we may someday come to possess the technology and competency essential to our pursuit of the past.

6.2. The Concept of Time in Human History and Prehistory

6.2.1. Culture, Language, Biology, and Time

It could be argued that time is among the most widely used and least understood concepts across all human groups. It is something everyone, everywhere is aware of and uses to structure their daily lives, though few of us understand what it actually is and how it works. We exist as slaves to its unstoppable and outwardly persistent advance, while also experiencing variable rates of its perceived passage, which is largely dictated by our chronological age, culture, and level of contentment with what we are doing at any given time. For instance, time seems to pass almost instantaneously while we are asleep or having conscious fun, but it can slog on assiduously when we are bored or brokenhearted.[156][157] Furthermore, time can feel as if it has rushed by, though somewhat paradoxically, when we look back at these events, they often feel as though they occurred a long time ago. Busy lives make time feel as if it passes quickly, while also making everything feel as if it occurred longer ago in the past. These dichotomies ardently indicate that the passage of time—as well as its tenacious forward momentum—is simply a product of our biological senses, imposing subjective rates of passage upon changes taking place in the physical world around us.

Despite time's ambiguity as both an abstract concept and physical component of the universe, it is important that we try to understand it, in some capacity, for several reasons. Not least among them is that our perception of time and the words we use to describe it can affect our general health and well-being. For example, a recent study investigating variation among linguistic groups with regard to the way they grammatically associate the present and future found that behaviors with a future-orientation tended to be associated with a better and more successful life than those behaviors that were not.[158] More specifically, individuals in cultures whose language forces them to grammatically dictate future events also have a tendency to save more money, practice safer sex, smoke less, and maintain a healthier diet, among other positive outcomes.[158] This indicates that time, and our perception of it, are important parts of our overall health, safety, and financial well-being.

Numerous cultures throughout the world have a somewhat restricted conceptualization of time, particularly regarding what may transpire in the future. For example, Native Americans of the Hopi Nation in the southwestern United States have very few words that refer to time. In fact, the closest future-specific words in their language are "sooner" and "later." This aspect of their limited lexis is generally attributed to a lack of verb tenses, which has the effect of curbing temporal cognizance, since most verbs describe actions that take place over time.[159]

The Pirahã tribe of Brazil also has a relatively loose relationship with time, which, like the Hopi, is mostly associated with the linguistic attributes of this cultural group. More specifically, the language of the Pirahã consists primarily of hums and whistles and contains no words that make explicit reference to time.[160] As a somewhat postmodern, ethnoscience interpretation of their perceived reality, this signifies that the passage of time and the way events are structured over the short and long-term are not an important part of their daily lives. The Pirahã people are unique in this regard, as well as for the fact that they have no descriptive words, they make no use of numbers or the subordinate clause, and they extend their kinship history back only two generations.[160] Furthermore, the Pirahã do not have any origin myths or intricate verbal accounts of past events, and they make no reference to their ancestors, which is very uncommon among human groups. These unique cultural characteristics are thought to be related to their general lack of broad-based temporal perception. Where instead of talking about what has already passed, or what is yet to come, they put emphasis on what happens "now," by living in every moment and experiencing it for all it is worth, because once it is over, it is over.[161]

As anthropologists, we have long considered this lack of temporal terminology in the context of the Sapir-Whorf Hypothesis, or what is also known as Linguistic Relativity, for its perceived parallels to Einstein's Theory of Special Relativity.[162] This concept was developed primarily by Benjamin Lee Whorf during the mid-20th century.[159] Whorf advocated that our language largely determines our thoughts and actions and that our culture-specific lexicon can dictate our way of perceiving, analyzing, and behaving in society and the environment around us.[163] Although anthropologists had largely abandoned this idea decades ago—at least in the strict deterministic sense—to some extent it can still inform our understanding of the cross-cultural concept of time.

For instance, a 2001 paper by Lera Boroditsky at Stanford University, titled *Does language shape thought?: Mandarin and English speakers' conceptions of time*, investigated the extent to which the language we speak affects how we think about the world.[164] Boroditsky's research showed that English and Mandarin speakers talk about time in different ways, which relates to how members of these distinct cultural and linguistic groups perceive the passage of time. The experimental and observational results of this broad-based study suggested that language is a vital factor in shaping patterns of thought, particularly with regard to how individuals and groups conceptualize time. However, Boroditsky also concluded that language was not a strict determinant of how we think and act, at least not in the causal Whorfian sense.[164]

Despite a lack of linguistic causality, it certainly makes sense that the words available to us play some role in influencing the way we perceive the world around us. However, while language can influence our thoughts and behaviors to some extent, words represent only a small part of the multitude of different ways we understand and communicate about the complexities of life. For instance, humans are capable of expressing ideas through visual art, which can be difficult to describe using verbal or written speech. The language of mathematics also provides a more in-depth means with which to communicate about aspects of space and time that exist beyond the limited number of words available to us in any language. Additionally, and perhaps most importantly, the words used in any linguistic group are a product of human cognition and perception, as much as they are a determining factor of them.

6.2.2. Time in the Mind's Eye

St. Augustine is credited with being the first person to deeply contemplate our complex relationship with time, having discussed its intrinsic nature as early as the 4th century AD.[165] Considering its many intricacies, Augustine was most interested in understanding how humans perceive observable experiences and how we can quantify the present using the concept of *now*, given that it has no actual duration. He pondered what is actually being measured when we enumerate time and how we gauge something that is durationless (such as the present), against something else that is also durationless (such as the past). In other words, when someone says an event took a long time, how did they quantify this thing that cannot be tangibly amassed? In this way, we infer time simply from relationships among different events, as an arbitrary thing measured in relation to other arbitrary things.

In essence, St. Augustine believed that time, and the way we perceive the passage and duration of events, is associated only with the mind. He postulated that our memory is what helps us quantify time, as it designates the more recognizable beginning and end of events in such a way that we are able to impose a subjective sense of finite duration upon them.[166] Impressively, Augustine's early views of how our memories and minds help us enumerate the fleeting present are still widely held today, more than 1,600 years after his initial temporal musings. In fact, according to Marcelo Gleiser, esteemed Professor of Physics and Astronomy at Dartmouth College,

> "Now" is not only a cognitive illusion but also a mathematical trick, related to how we define space and time quantitatively. One way of seeing this is to recognize that the notion of "present," as sandwiched between past and future, is simply a useful hoax. After all, if the present is a moment in time without duration, it can't exist. What does exist is the recent memory of the immediate past and the expectation of the near future. We link past and future through the conceptual notion of a present, of "now." But all that we have is the accumulated memory of the past—stored in biological or various recording devices—and the expectation of the future.[167]

Our biological perception of the rate at which time passes, and the idea of a durationless *now*, also perplexed Einstein, who confessed in a conversation with the philosopher Rudolf Carnap that there is "something essential about the now." However, Einstein conceded that whatever "now" is, it lies "just outside the realm of science."[168]

St. Augustine's focus on the mind as a quantifier of time certainly makes sense in the context of how we conceptualize its passage throughout life and in association with the perceived duration of specific events. For example, differences in how we remember time passing are apparent in thinking back on different periods of life and how it seems to pass more quickly as we grow older. It is also palpable when hanging out with young children for any amount of time, as they perceive time, both in a moment-by-moment and long-term capacity, very differently than an aged adult, who experiences a continual thinning out of time throughout life.[169]

These aspects of our differential perception of time, and how our memories and cognition shape its arbitrarily derived duration, also extend to other conscious animals. Collectively, this suggests a biological form of temporal relativity that is woven into the physical fabric of universal spacetime. For instance, our perception of the differential rate of activity among hummingbirds, ants, barn swallows, elephants, sloths, etc., is an important reminder of just how much the perception of time is in the mind's eye, while also being highly integrated with the size, age, metabolic rate, predator/prey status, and selective fitness of any organism.[170]

Additionally, there is no reason to assume other organisms perceive time any faster or slower than us humans, despite their divergent rate of activity relative to our anthropocentric speed of life. Rather, a hummingbird likely perceives time at the same relative speed as us, but simply sees us, and most other organisms, moving much more slowly by comparison. Fast-paced animals such as birds and flies experience each passing moment in slow motion relative to our reference frame, though at an adaptively adjusted speed that favors their survival, as determined by subsistence and predator-avoidance needs throughout their long evolutionary history.[171]

This variability in temporal acuity within and among different animals may be best described as *Biorelativity*. As it too is analogous to Einstein's Special Relativity, but among living organisms capable of perceiving events at variable rates relative to other living creatures, each with different speeds of cognition and biometric characteristics. Unfortunately, various challenges currently limit our ability to research the speed at which living organisms cognitively perceive the passage of time, which includes the vexing question of how to quantify the duration of *now*. Although in spite of these challenges, we continue to peel back the layers of opaque complexity that circumscribe time—in the physical, biological, cerebral and cognitive sense—as we draw nearer to a more holistic understanding of that which is so familiar, yet still so unknown.

6.2.3. The Measurement We Take of Change

Many long-term changes have occurred throughout the long history of the universe. To humans here on Earth, these changes appear to have taken place over what we designate as billions of years, based upon units we derived from the daily rotation of our Earth and its yearly rotation around our sun. However, relative to another observer, enduring celestial and geologic events could be perceived as having occurred in a fraction of our earth-devised time or, potentially, all at once. This aspect of temporal relativity elicits questions about whether time exists as a tangible physical entity, or merely as something we have conjured up in order to make sense of and to measure the perceived changes to the state of matter around us.

We can appreciate that time has passed, simply by observing matter change from one state to another, or moving from one location to another, regardless of the "speed" at which these changes appear to have occurred relative to any one observer. We are also accustomed to time moving in only one direction, despite the fact that nearly all other aspects of the universe are symmetrical and cyclic. We bear witness to and quantify time's passage in association with an established and predictable order of events, derived from our own past experience, in which earlier causes lead to later effects.

Regardless of whether there are conscious beings that believe in time, who possess the necessary senses to perceive it, or who have developed the technology to record it, changes still occur in the natural universe. This indicates that time is indeed a real phenomenon, no matter how or at what speed it passes relative to anyone's or anything's frame of reference. However, because "now" is indefinable, it is possible—and most physicists actually agree—that all time exists all at once. In fact, it may only be when conscious beings are present that time appears to move at all, and in order to make sense of the world around them, always with rectilinear motion.

This conventional view of the universe considers the three dimensions of space and one dimension of time to exist everywhere and all at once, as one massive block of spacetime. In other words, all events that lie to our past, as well as those to our future, have already occurred. Events occurring now and in our immediate past are clearly more tangible, as we have access to memories of them. However, future events remain sheltered from us, until which time as we exist alongside them, as part of our collective forthcoming consciousness. In the context of physical universal time, we are already there at those points in the future, as well as everywhere else we have ever been during every moment of our lives. However, because of biological constraints that only allow us to see time unfold linearly, we are only able to access memories of moments that make up our past.

This all-pervading interpretation of universal spacetime is widely recognized among physicists, philosophers, and mathematicians, and is commonly referred to as *Landscape Time or Block Time*. An implication of this landscape view of time is that if everything that was, is, and ever will be already exists, then all that remains is how we, as sentient beings, observe events that have already taken place as part of our future, up until the day we die. This complex issue of how consciousness relates to the structure of time and the physical universe was addressed by philosopher and mathematician Hao Wang in the article *Time in philosophy and in physics: From Kant and Einstein to Gödel.*

> When I project my consciousness of time onto the world, I get an idea of objective or absolute time. On the one hand, there is a continuing sequence of world-states which individually are at rest and collectively constitute the material content of time. On the other hand, there is a mysterious process of lapse or flow or change, by which every world state travels through being future, present, and past. If we leave out my consciousness and that of other beings, then it is hard to see what is so special about Now, or indeed to make sense of the very distinction between past, present and future. Without this distinction, however, time would be like space in the sense that there is no flow and there is no distinguished direction or arrow of time.[172]

In his book *Time Journeys: A Search for Cosmic Destiny and Meaning*, Paul Halpern, a physicist at the University of the Sciences in Philadelphia, also examined this question of how consciousness relates to block time. Although, he takes it a bit further by noting that if our perceived notion of linear time is a product of our biological consciousness, then reversing this may help us figure out how to reverse the obstinate arrow of time.

> Perhaps all of space and time exist at once and our travels through time are simply something that our conscious minds undertake. If we could break this force that is propelling us forward, maybe we could travel back in time.[173]

It is certainly understandable why a consistent and shared linear perception of time would develop among sentient beings, considering that a categorical separation of *nows* may be necessary for making sense of and surviving in the environment around us. It is also fathomable how a highly advanced sentient being that has devised ways of gauging the conscious and physical properties of time could use that intellect and advanced knowledge to reverse its direction.

In its simplest form, among human and non-human animals alike, time can be thought of as the quantification of sequential events, relative to whomever or whatever is doing the gauging.[174] If no change occurs, then there can be no time. If the sun always stayed in the same position in the sky, the seasons never changed, no one was ever born or died, blinked or waved, then there would be no way to measure time, as there would be nothing for it to be relative to. However, without any change, matter could never take on new forms, and there would never be anyone to blink, wave, or die in the first place. So naturally, as we exist *now*, so does time, and so do the changes that have, are, and will take place across the whole of block time, as our consciousness continues to catalog each state giving rise to the next, in perpetuity, for the duration of life's time.

6.2.4. Our Measurement of Time…Through Time

Various societies throughout human history and prehistory have devised numerous means of quantifying time, with most using regularly occurring events as a metric with which to judge the passage of time. The earliest and most temporally and spatially ubiquitous of these tended to be celestial, such as the cycles of the moon or the sun rising and falling in the sky. These were later expanded upon using mechanical devices, such as sundials, hourglasses, and *clepsydras*, which measure time by the regulated flow of liquid (most often water), into or out of a vessel. This type of water clock was developed as early as 6,000 years ago in China, and around 3,500 years ago in Babylon and Egypt.[175]

More recently, or at least since the mid-20th century, the majority of the world has based its collective consensus of what time it is using cesium and rubidium atom clocks. These modern clocks are so reliable that they are able to "keep time" with an accuracy of 2 nanoseconds per day, which is the equivalent of losing only one second over the course of 1,400,000 years.[176] Our increasingly advanced ability to measure completely arbitrary time on a progressively more precise scale, has also provided us with a deeper understanding of space, time, light, and gravity.

More specifically, advances in the quantification of time have allowed us to demonstrate that it passes at variable rates among people (and clocks) moving at different speeds in different reference frames as noted above. It has also helped demonstrate that gravity, including that of Earth, elicits a similar effect as high-speed motion, altering the passage of time between different observers situated near different gravitational masses. Einstein's 1905 paper, *On the Electrodynamics of Moving Bodies*, demonstrated that space and time are relative to the observer.[177] Then, ten years later, he showed how gravity and the curvature of spacetime can elicit these same time dilation effects.[178] [179] [180] Together, these discoveries had a tremendous impact on our understanding of time and the universe.

> In 1905, Einstein changed altogether our notion of time. Time flowed at different rates for different observers, and Minkowski, three years later, formally united the parameters of time and space, giving rise to the notion of a four-dimensional entity, spacetime.[181]

This referential reality of spacetime, developed by Einstein and expanded on by his former mathematics professor Hermann Minkowski, has since been experimentally verified many times over. One of the earliest experiments that helped demonstrate this mind-bending time dilation effect took place 66 years after Einstein initially proposed it. In October of 1971, J.C. Hafele of Washington University in St. Louis and Richard Keating from the U.S. Naval Observatory procured four cesium beam atomic clocks from the Naval Observatory in hopes of testing Einstein's 1905 hypothesis.

Based on Einstein's theory of relativity, clocks moving at different speeds relative to one another should read different times at the end of their journey. To test this, Hafele and Keating kept two clocks at the Naval Observatory and loaded the others aboard commercial airliners to fly them around the world; initially from west to east, and then again from east to west. Although commercial airliners fly less than one millionth the speed of light, the precision of these cesium atom clocks was enough to detect the very small effects of time dilation at these very slow speeds, relatively speaking. The results, published in the journal *Science* in 1972, showed that during the eastward journey, the clocks ran an average of 59 ± 10 nanoseconds (billionths of a second) slower than the atomic clocks kept at the Naval Observatory. By contrast, on the westward journey, the clocks ran an average of 273 ± 7 nanoseconds *faster* than the earthbound clocks.[182] So not only did these results corroborate Einstein's predictions regarding the effects of high-speed motion on clocks and the passage of time, but they also confirmed another prediction he made…

Einstein predicted that the rotational speed of the Earth also produces a kinematic time dilation effect, meaning a slowing of time irrespective of Earth's mass and the influence of gravitational time dilation. This was clearly indicated by the difference between Hafele and Keating's clocks after completing their westward vs. eastward journeys, despite the fact that both planes were flying at the same speed in each voyage around the Earth. More specifically, the clock aboard the eastbound plane was moving in the direction of Earth's rotation, which meant that it had a greater velocity, resulting in a time loss relative to the clocks "at rest" on the ground. However, the clock aboard the westbound plane was moving against the Earth's rotation, which meant that it actually had a lower velocity than the clocks on the ground, causing it to tick slightly faster by comparison. This pioneering study experimentally demonstrated that the rate of time's passage varies among clocks in different inertial and gravitational reference frames, which helped support Einstein's theories of special and general relativity.

As we continue to evolve the technology we use for measuring time, Einstein's predictions continue to be upheld. For example, using optical atomic clocks, which are considered to be at the forefront of modern measurement science,[183] researchers recently demonstrated that time dilation is observable at speeds and gravity potentials far lower than what was observed in the Hafele/Keating experiment. Using a 75-meter-long optical fiber connecting two optical atomic clocks, researchers were able to observe kinematic time dilation occurring at speeds as low as 10 meters per second, as well as gravitational time dilation at a height of only 1 meter above the Earth's surface.[184]

As a prominent researcher of time, Einstein was frequently asked to explain what time is, to which he would often respond "time is what clocks measure." While this may seem like an attempt to dodge the question, this simple response highlights the subjectivity of time, as well as the apparent disconnect between the biological perception and mechanical properties of it. Additionally, this operational view of time helped Einstein move past the prevailing Newtonian view of absolute time, in favor of his new relativistic interpretation, centered on motion and the force of gravity.

According to Dr. William Phillips, a Nobel Prize award-winning physicist at The National Institute of Standards and Technology whose work has greatly contributed to the creation of better atomic clocks:

By taking seriously the idea that time is what a clock measures
Einstein was able to come up with a deeper understanding of the nature of time than had been the case before. In particular, he thought
about a particular kind of clock and he imagined what would happen
if this clock were moving with respect to an observer....Taking the
idea seriously, that time is what a clock measures, you come to the
conclusion that time is running more slowly for the person who is
moving from the observer's point of view. We now know, from the
basis of experiments, that this is true.[185]

Einstein's research helped demonstrate that all clocks are equally affected by motion and gravity, including biological clocks within the brains of
conscious beings like ourselves. In other words, observers with their own
set of internal and external clocks experience different intervals of time
between the same two events, in accordance with their relative speed or
the amount of gravity in their respective reference frames.[186] Advances
in our ability to precisely measure the passage of time using increasingly
sophisticated clocks has helped frame our perception of time as sentient
beings, while also offering up a more holistic understanding of spacetime
as a whole. However, the question still remains, why does time seem to
march incessantly forward, regardless of how fast or slow it appears to
pass for any observer, anywhere, and at any time?

6.3. The Arrow of Entropy and the Arrow of Time

Life's moments seem to follow a linear path through time. We observe
and are taught from a young age to believe that an initial cause elicits
a subsequent effect. Many pre-schools and kindergartens provide kids
with worksheets requiring them to put events in sequential order, further
solidifying our culture-specific notion of linear time during an important
stage of cognitive development. Although, somewhat paradoxically, most
of the things we use to measure time are cyclic in nature, such as the hands
of a clock, the daily rotation of the Earth around its axis, and the Earth's
yearly rotation around the sun.

There are many cultures around the world that do not adhere to this
staunch rectilinear interpretation of time. As mentioned above, some pay
little attention to time, not speaking of it and presumably not ascribing
any meaning to this concept at all. Many others view time as flexible,
meaning that they are reluctant to measure it or to try to control it, while
others actually embrace a more cyclic view of time where we are thought
to exist within the patterned reoccurring processes of nature, adjusting

to the cyclic, repetitive patterns of life, within the greater realm of the natural world.[187]

Regardless of our cultural interpretations of time, there is an undeniable sense that a cause always precedes an effect and not the other way around. For instance, an egg falls off the counter and breaks on the floor; however, a disorderly mess of shell, yolk, and albumen never lifts off of the floor and forms an amalgamated egg on the counter top. This advancing order of events structures our conventional understanding of linear time and is best understood in the context of entropy and the second law of thermodynamics.

Since the big bang, all matter in the universe has been transitioning from a state of order to one of disorder and randomness, and the ostensible direction of the arrow of time is intricately linked to this arrow of entropy. Curiously, the enduring process of order decaying into disorder does not reverse itself, as entropy cannot decrease within isolated thermodynamic systems. This makes entropy the only thing known to the physical sciences that requires a specific direction of time and helps explain why certain processes seem to occur spontaneously, while their time reversals do not, even though the conservation laws of physics allow for both. According to Thomas Kitching, astrophysics lecturer at University College London:

> While we take for granted that time has a given direction, physicists don't: most natural laws are "time reversible" which means they would work just as well if time was defined as running backwards. ...
> In the dimension of space, you can move forwards and backwards; commuters experience this every day. But time is different, it has a direction, you always move forward, never in reverse. So why is the dimension of time irreversible? This is one of the major unsolved problems in physics.[188]

6.3.1. Entropy and Biology

The asymmetry that characterizes entropy and time also applies to biological organisms, despite the fact that life is an entropy reducer, meaning that it is capable of creating order out of disorder and reducing entropy in its localized environment. However, on the whole, life still increases the overall entropy of its surroundings as well as the entire closed system of the universe.[189] Consider the consumption and breakdown of food, for example. The body takes organized fuel in the form of food, breaks it down into a less organized state, then produces heat and chemical energy that increases entropy in the organism's proximate environment. In fact, even while sitting completely still, we are continually dissipating heat energy, which acts to increase the overall entropy of our surroundings. Additionally, when a living thing dies, it tends toward chaos through the mechanism of decay, a process by which organisms trade the energy of past life for that of future life, as the entire biome drives persistently forward along the arrow of entropy and the arrow of time.

Jeremy England, an assistant professor of physics at the Massachusetts Institute of Technology in the United States, has suggested that entropy might have been a prime mover behind the origins and evolution of life, in which the earliest lifeforms were driven to increase entropy production over time.[190] In a 2013 paper in the *Journal of Chemical Physics*, England presented a mathematical formula showing that the most successful evolutionary pathways also tended to be those that were better at capturing energy from the environment and dissipating it as heat. In this model, living beings could have evolved from mere inanimate clumps of atoms exposed to an energy source, such as the sun or some type of chemical fuel, while surrounded by a heat-bath, such as an atmosphere or an ocean. This would generate simple, but increasingly complex lifeforms, which were inclined to reorganize themselves in order to dissipate more energy, more effectively over time.[190 191 192] In this way, life's origins may be associated with a dissipation-driven adaptation of matter, in which the physical properties of the environment encourage organic elements of that environment to restructure themselves to dissipate greater amounts of energy and with greater efficiency, across each generation.

An important implication of this model is that life, as well as its tendency to self-replicate, may be inevitable when certain basal environmental conditions are in place. After all, there are few better ways to dissipate greater amounts of energy over time than to create copies of oneself. For instance, a plant is far better at capturing and dissipating energy back into the environment than a group of simple carbon atoms, and more plants (generated through sexual or asexual reproduction) can capture and dissipate exponentially higher amounts of energy, as they continue to reproduce themselves into the future. Add to this evolutionary advancement an intelligent species capable of producing entropy machines and it is easy to understand why global warming and climate change have become so concerning in recent times.

6.3.2. Entropy as a Measure of Time

Based on these principles of dissipative adaptation and driven self-assembly, we can conceptualize how entropy may have contributed to the origin and evolution of life. As the most cognizant species on this planet, humans are also able to observe and mathematically demonstrate an arrow of entropy and, by association, an arrow of time. This aspect of the world around us is not only tangible and an integral part of our everyday experiences, but it is also all we have ever known. Our perception of time is based on a comfortable and well-established notion that a broken egg will not spontaneously reform, an airplane will not reverse its direction in the sky, and we will not suddenly begin to get younger. Though because entropy is the only measure that requires unidirectional time—while nearly all other microcosmic physical processes are believed to be time-symmetric—it raises an important question about whether entropy is time-based or whether time is entropy-based.

We often think of entropy as moving time forward. Although, the concept of entropy is also a product of our perception of changes occurring with one thing relative to another, which makes it a consequence of time as much as it is a driving force behind it. In this way, entropy may be just another way we enumerate time, like our sundials, hourglasses, clepsydras, and atomic clocks. We could be discovering the true nature of time through the study of entropy and the second law of thermodynamics. However, we could also just be learning about our interactions with the world in a much more metaphysical sense, generating a mathematical understanding of time based solely on what we are capable of observing in our narrow region of space and as part of our short duration as an intelligent species, relative to the vast entirety of four-dimensional spacetime.

6.3.3. Future Cause → Past Effect—Wheeler's Delayed-Choice Gedanken Experiment

An increase in entropy accompanies every real process observable to us. Although, this does not necessarily mean that these changes are what moves time in a forward direction. Rather, they may simply reflect a different conditional state that exists in a different slice of block time, which, outside of our biological frame of reference, may in fact be time-symmetric. Nevertheless, we perceive, and have come to accept, that an earlier cause gives way to a later effect. However, recent research in quantum mechanics may challenge this widely held notion by offering evidence that the arrow of time is simply a biologically derived illusion, which masks far more complex interactions among matter. This was recently revealed by a team of researchers at the Australian National University's Research School of Physics and Engineering. Most notably, they demonstrated that the future condition of a photon can affect its past state, as we view them relative to one another in time.[193] In their 2015 article in the journal *Nature Physics*, these researchers provided evidence in support of Wheeler's delayed-choice gedanken experiment, made famous by renowned physicist John Archibald Wheeler. This was noteworthy, because prior to their study the delayed-choice gedanken experiment was considered to only be a thought experiment (gedanken means "thought" in German), since no one believed it could actually be tested in reality.

Wheeler's original idea was that a photon, which can act as a particle or a wave, may actually "decide" which one it is going to be, based on the future condition that is presented to it by the scientists attempting to study it. This would indicate that simply observing its behavior acts to help the photon decide which it is going to be, a particle or a wave. So in this context, observation defines reality, and future circumstances are capable of influencing those in the past.[194]

It is well established that light has properties of both a wave and a particle, which is aptly known as wave-particle duality. This was first demonstrated by the physicist Thomas Young in 1801 when he carried out what was to become a common research procedure known as a double-slit experiment. In this simple experiment, light is emitted in the direction of a screen with two narrow vertical slits. Photons radiating toward the screen can be seen acting as a particle, where they cast a glow as two vertical stripes on the wall behind the screen. Or, they can act as a wave, interfering with each other and scattering the light, which shows up as multiple vertical stripes on the wall behind the screen.

What is most unusual is that when researchers attempt to observe a photon during these experiments, it adopts a definitive discrete state, as either a wave or a particle, as if the photon is making a cognitive decision based on the researcher's parameters for observation. The genius of Wheeler's 1978 delayed-choice thought experiment was that if it were possible to add a second screen behind the first, researchers should be able to determine *when* that photon "chooses" to act as a particle or wave.[194] This second screen is vital to the experiment because once it is inserted, it would cause interference in the same manner as the first screen. This would allow the experimenter to observe the state of the photon after passing through the first screen and see if it remained the same after passing through the randomly inserted second screen.

It was believed that Wheeler's experiment could never be carried out because the experimenter would have to place the second screen behind the first, in the path of the photon, at random, and just after the photon had already passed through the first screen. Because of the tremendous speed at which light travels, this scenario leaves a very short window with which to accurately and reliably insert the second screen. However, the research team at the Australian National University were able to conduct a version of Wheeler's gedanken experiment using a helium atom as a surrogate for the photon and lasers to act as the arbitrarily inserted second screen.[193] Their results consistently showed that when both of the two screens were in place, the helium atoms acted like waves, passing through the screens along many different paths. However, when the second screen was not inserted, the atoms passed through the first screen as particles. Remarkably, the helium atoms appear to have "decided" to pass through the first screen as a wave, even *before* the second screen had been inserted. Additionally, the atoms were already acting like particles during random trials in which the second screen was not inserted, and again, before it was ever determined if that second screen would be present or absent.[193]

These results indicate that the future state of the second screen, as either present or not, has an effect on whether the photons leave the first screen as a particle or a wave. In essence, the future event (cause) would appear to dictate the past state (effect) of the helium atom, as if it somehow knew ahead of time whether the second screen was going to be inserted. This stands in stark contrast to how we perceive cause and effect in a linear conceptualization of time and indicates that our notion of time flowing only from past to future is largely a product of our perception. It also suggests that free will is an illusion and that the measurements we take while observing natural phenomena can affect, or at least be highly integrated with, a much more complex predetermined reality.

6.4. Block Time and the Opacity of Free Will

We tend to view time as changes that can be measured in relation to other things that occur in patterned ways. This helps us define what time is and frames it in the context of a linear process, in which one state gives rise to a slightly modified version of what came before. However, this recent test of Wheeler's delayed-choice gedanken experiment, as well as other modern physics research, suggests that certain future events might also influence past events. This indicates that our conventional perception of linear time may be an illusion likely rooted in our biologically derived consciousness.

The past and future are both quantifiable, though we lack any concrete means with which to gauge the durationless present, which points to the nonexistence of "now." As Einstein once said, "People like us who believe in physics know that the distinction between the past, the present, and the future is only a stubbornly persistent illusion."[195] Indeed, the most widely held notion of time among physicists is that all matter in the universe exists as part of an immense 4-dimensional block of spacetime, where the past, present, and future state of that matter exist together, without any clearly defined present.

Interestingly, at any point in our lives—along our *world line* as it is known to physicists—we exist as slightly different versions of ourselves, which often occupy the exact same spaces, over and over again, albeit at different times. Extrapolated further, our Australopithecine ancestors and presumably our future extratempestrial descendants are all walking around on Earth right *now*, just in separate slices of this omnipresent timescape. According to Adam Frank, professor of theoretical astrophysics at the University of Rochester:

> The remarkable thing about space-time is that it contains all the events that ever happened. It also includes all the events that ever will happen. So, in this big, 4-D space-time representation of the universe, there's Julius Caesar getting stabbed and the Mets winning the 1969 World Series. But the coffee you are going to spill on your pants at that meeting next Tuesday is there, too … In fact, everything that will ever happen to you—including your death—is strung across space-time as a linked string of *already existing* events. Physicists call this your world line.[196]

An important implication of block time is that if everything that was, is, and ever will be already exists, then free will as we known it cannot. It undoubtedly feels as though we are able to alter the course of events by making decisions, which then manifest themselves as some future outcome. However, once our consciousness is positioned along our world line in such a way that we are able to look back through life—and view our previously unknown future as a completed past—it is much easier to see how there was only ever one outcome, regardless of how much we felt like we were free to decide along the way.

After spending many years researching the perception of free will, particularly in the context of block time, it became clear that this debate can be boiled down to a forward-looking vs. a past-looking perspective. In other words, looking toward the future, it seems as though there are countless decisions we could make that would lead us down countless different paths through life, which bolsters our sense of free will. However, once we are a part of that future—looking back at our lives as a multitude of integrated events in the past—it is clear that only one path was ever actually followed, and only one outcome was derived from what appeared to be an endless set of possibilities. Furthermore, there is no evidence to suggest that any other outcome ever existed, here or anywhere else.

On a larger scale, taking into account the whole of human time, we must also consider that our unknown future already exists as an observable past to our distant descendants. Moreover, once we arrive in that future and become able to look back at this past, so too shall we see a lone history, a single path meandering toward a predestined present. Irrespective of where we are in the lifelong process of turning the unknown into retrievable memories, there is only ever going to be one reality in the end. According to renowned British physicist Paul Davies, in his book *About Time: Einstein's unfinished revolution*:

> Human life revolves around the division of time into past, present and future; people will not relinquish these categories just because physicists say they are discredited . . . This is perhaps what disturbs people most about block time. If the future is somehow "already there," then we can have no hand in shaping it . . . In their professional lives most physicists accept without question the concept of the timescape, but away from work they act like everybody else, basing their thoughts and actions on the assumption of a moving present moment.[197]

This aspect of block time, which challenges our entrenched notion of free will, tends to make people very angry and combative, likely because we want to feel as though the decisions we make are manifested in some sort of chosen outcome. Though as Davies writes, even if free will doesn't exist, it really doesn't change anything about our day to day lives, given that we still *feel* as though we are making decisions that will elicit a later effect, even if it is already part of a predestined future.

6.4.1. Feeling Freely Willed

The false sense that we are making decisions remains an important component of our lives. It is also a common aspect of ordinary conversation, as we constantly refer to future events as if we were somehow dictating their outcome. What do you want to do today? What should we have for dinner? What time should we leave for the show? Conversations like these are perceived as shaping later events, but they actually only exist as an integral part of what was always already going to happen. We are constantly "planning" the future, in the short and long-term, despite the fact that it is already determined to be the way it always was. However, if we didn't talk about the future, and if we didn't do the things we had talked about—whether they work out the way we planned or not—we would never arrive in the place and time we were always going to be.

If it wasn't disheartening enough that all of our decisions are predetermined, a growing body of research in the fields of neuroscience and psychology also indicate that we may not even be a part of the decisions we feel we make. Rather, these conscious "decisions" actually occur *after* the neurological triggers that initiate the specific action. In other words, we become aware of what is happening after our brains have already begun the necessary cognitive functions to carry it out, despite feeling like we are the cause of these actions. One of the earliest, and perhaps most influential of these studies, was conducted by Benjamin Libet and his colleagues in the early 1980s.[198]

Published in the journal *Brain* and cited over 2,830 times, their experiment required research subjects to flex the finger or wrist of their right hand, whenever they felt like doing so, while their muscle and brain activity were monitored by an electromyogram (EMG) and electroencephalogram (EEG). Additionally, the research subjects were told to report the exact moment that they decided to move their hand. The results of this study provided experimental evidence that the conscious intention to act comes after the onset of the cerebral activity associated with that action.[198] While it had already been established that brain activity can be detected up to a whole second before a motor response, what was interesting about the results of Libet's study was that the brain activity responsible for the finger movements occurred about 300 milliseconds before the test subject reported an intention to move their finger. This suggests that our brains decide to initiate an activity before we have any actual awareness of that decision.

As stated previously, there are numerous methodological difficulties involved with attempting to measure the perception of time among biological organisms. As such, some have criticized Libet's procedures for measuring the time at which his research subjects felt the urge to move their fingers. However, subsequent studies, which used different methods for measuring this important variable, have consistently corroborated Libet's results.[199] This includes a recent single-neuron study, in which neuronal recruitment was found to occur upwards of 1,500 milliseconds before the research subjects reported "making the decision" to move their fingers.[200]

Other studies have also shown that the brain is capable of binding intentional actions to the effects of those actions, and manipulating the order of these events to make us feel as if we have initiated the response, even though it occurred before we become aware of it.[201] This phenomenon may help explain why measured brain activity and the subjective experience of things being done to us are far more intense than when we do them to ourselves (e.g., being tickled by someone vs. tickling ourselves).[202 203]

Perhaps the most interesting takeaway of these initiation-perception studies is that even our most basic activities appear to be predetermined, in that our brain initiates an action even before we are consciously aware of the urge to perform it. This aspect of the illusion of free will is summed up well by Chris Frith, Professor Emeritus of Neuroimaging at University College London:

> The implication of this observation is that, by measuring your brain activity, I can know that you're going to have the urge to lift your finger before you know it yourself....We think we are making a choice when, in fact, our brain has already made the choice. Our experience of making a choice at that moment is therefore an illusion. And if we are deluded in thinking that we are making choices, then we are also deluded in thinking that we have free will.[204]

6.5. Block time, Backward Time Travel, and Self-Consistency

Another important implication of block time and the omnipresence of space and time is that nothing and nobody can ever "change" anything in the past or future. We are simply in the act of becoming a part of what already is, regardless of whether we are looking from the past toward the future or from the future back toward the past. In spite of a persistent perception of innumerable alternative outcomes, the past and future cannot be altered in any way, given that the relationship among all events within them are already established.

To say that the past has been altered, one must first assume that there was an original course people in the past were expected to follow en route to the future. A problem with this assumption, however, is there is no evidence of any alternative histories ever existing. We like to think that there could have been many different outcomes in our lives if we had just "made a different decision" or "done something differently." Though in the end there was only one outcome and no indication that it ever could have been any other way.

Block time is also useful for conceptualizing how we may someday change our position within it and what might happen once we do. For instance, if time exists as a physical component of the universe, then in the same way we are able to accurately calculate the proper trajectory through space to land a rover on Mars, or the Philae lander on comet 67P, so too may we be able to calculate the proper trajectory through the matrix of predetermined events across 4-dimensional spacetime, so as to accurately arrive at a previously existing point of it.

This one-history elucidation of spacetime is a vital part of understanding the feasibility of backward time travel, as it provides clarity to a circumstance that many consider to be overly complex and riddled with paradox. For instance, some individuals have postulated that interjecting oneself into the past would cause a ripple in the fabric of spacetime, which would "alter" history and therefore change subsequent future events. Others have suggested that this would cause the time traveler—or possibly the entire universe—to enter a different dimension, with a different set of future outcomes determined by whatever changes the time traveler made to whatever past was thought to have existed before. Yet others have proposed that changes made to the past would impact only that one single point in time, and the universe could somehow correct itself and eventually return to same predetermined future reality, as if it never happened.

These views on the effects of backward time travel are mostly derived from an anthropocentric future-looking perspective, with misconceived notions of free will and a false sense that we could ever change anything about the past or future. Conversely, the growing consensus among physicists is that any instance of intertemporal interaction always existed as an integral part of those periods of time. In other words, events associated with someone from the future interjecting themselves into the past had always already been a part of that past. Furthermore, the time traveler—whether it felt like free will or not—was simply doing what she/he was already going to do while visiting that past time. Consequently, there are no paradoxes associated with this trip through time, as everything—regardless of whether it is viewed from the past or the future—remains self-consistent throughout.

6.5.1. *The Novikov Self-Consistency Principle, Revisited*

As mentioned at the start of this chapter, the idea that an inherent self-consistency exists among events connected by bridges through space-time is known as the Novikov Self-Consistency Principle. Originally introduced by Russian physicist Igor Dmitriyevich Novikov and developed in collaboration with Yakov Borisovich Zel'dovich, among other prominent physicists of the 1970s, 1980s, and beyond, the self-consistency principle has been instrumental in providing clarity regarding a number of problems thought to arise as a result of visiting the past.[205] [206] [207] Additionally, in a 1990 paper titled *Cauchy Problem in Spacetimes with Closed Timelike Curves*, Novikov and his colleagues even began to arrive at some consensus, furthering the notion of an intrinsic self-consistency among events linked by CTCs, even with regard to the most enigmatic of all paradoxes, killing one's younger self:

> The only type of causality violation that the authors would find unacceptable is that embodied in the science-fiction concept of going backward in time and killing one's younger self ("changing the past"). Some years ago one of us (Novikov[207]) briefly considered the possibility that CTCs might exist and argued that they cannot entail this type of causality violation: Events on a CTC are already guaranteed to be self-consistent, Novikov argued; they influence each other around a closed curve in a self-adjusted, cyclical, self-consistent way. The other authors recently have arrived at the same viewpoint. We shall embody this viewpoint in a *principle of self-consistency,* which states that *the only solutions to the laws of physics that can occur locally in the real Universe are those which are globally self-consistent.*[148]

6.5.2. The Grandfather "Paradox"

Among the most well-known of all backward time travel paradoxes is the *Grandfather Paradox*, which is part of a broader category of enigmas known as *consistency paradoxes*.[183] This hypothetical scenario involves a person going back in time and killing their grandfather, at a time that predates when their parent was born. The paradox arises because an individual should not be able to exist, let alone go back in time and kill their grandfather, if one of their parents was never born. While tantalizing, numerous studies have consistently found that self-consistency is maintained in any backward time travel scenario, including the grandfather paradox, which effectively sidesteps any absurdities that would seem to arise if one were to attempt to "alter" the past in any way.

A number of studies in the early 1990s attempted to simplify and resolve this paradoxical grandfather-murder scenario, by conceptualizing a time machine with two mouths, labeled A and B. In this situation, when a time traveler (represented by a billiard ball) enters mouth B they will be transported to the past via mouth A, and at a time just before it entered mouth B of the time machine. In the context of the grandfather paradox, as the billiard ball is about to enter mouth B to return to the past, the older version of itself—which had entered the time machine previously—comes out of mouth A and changes its trajectory, so that the billiard ball time traveler is never able to enter mouth B in the first place. This presents a paradox, given that the older version of the billiard ball should never have been able to enter the time machine, if it knocked itself away from the time machine before it ever went in.[148 208 209 210]

This is analogous to the grandfather paradox, because a person should not be able to kill their grandfather if one of the parents essential to making them never existed. However, as Novikov points out in his book the *River of Time*, "If a 'time loop' exists, the events on this loop cannot be separated into future and past."[211] So even though these billiard ball and grandfather-killing scenarios seem paradoxical, they cease to be problematic once we take into account the effect of this collision from the beginning of the model, rather from the end. For instance, if the original younger ball is rolling on a path that does not lead to mouth B of the time machine, but its older self comes out of mouth A and redirects its younger self into mouth B, then there is no paradox. Also, after hitting the younger ball and knocking it into the time machine, the older ball that emerged from mouth A in the past, will continue rolling in a direction that does not lead to mouth B, but does create the initial motion that sets it up to be hit directly into mouth B of the time machine, and so it goes. Perhaps most importantly, as Novikov points out, this type of self-consistency resolution appears to happen naturally, thereby preventing paradoxes from ever presenting themselves at all.[211]

In fact, according to Novikov, no known proof exists in which paradoxes arise in this or any other billiard ball scenario like it.[212] This critical assessment was supported by physicist Kip Thorne, who—after lengthy discussions with another prominent researcher in this area named John Friedman—concluded that backward time travel does not necessarily result in paradoxes, regardless of what or who is sent through the time machine.[213] Furthermore, while Novikov, Thorne, Friedman, Polchinski, Echeverria, Klinkhammer and others were conducting this research, David Elieser Deutsch, a British physicist at the University of Oxford, was investigating similar problems using quantum mechanics. Deutsch's research also corroborated these billiard ball time machine studies, as he demonstrated that any paradoxes created by CTCs could be avoided at the quantum level, simply because of the way fundamental particles behave.[149]

In Deutsch's scenario, the CTC leads back to a machine where a particle flips a switch. Upon being flipped, the switch produces a particle that will go back through the CTC and flip the switch again…or it will not, and the machine will emit nothing. This solution involves no deterministic formation of the particle, but only the chance that the switch will or will not be flipped by the particle, which has a ½ probability of entering the CTC, emerging from it, and flipping the switch. Deutsch's solution showed that any particle that enters a CTC must emerge at the other end with the same exact properties as it entered. So a particle produced by this machine, which was created with a probability of ½, would enter the closed timelike curve and emerge from the other side to flip the switch with the same ½ probability as when it entered.[149] This scenario can also be extrapolated as a quantum solution to the grandfather paradox. For instance, say an individual is born with a ½ probability of going back in time and killing their grandfather. In turn, the grandfather has a ½ probability of escaping the grandchild's attempt at murdering him. Because all of these probabilities are ½, and not a number that guarantees that the event will happen, probability says that this is no longer a paradox, as it is not absolute.[214]

More recently, at the University of Queensland in Australia, physicist Tim Ralph and his PhD student Martin Ringbauer, led a team of researchers who were able to simulate Deutsch's model experimentally for the first time. In their 2014 *Nature Communications* paper, titled *Experimental simulation of closed timelike curves*, they confirmed many aspects of Deutsch's 20-year-old theory, particularly how it relates to the classic grandfather paradox.[150] More specifically, Deutsch's theory of probabilities and particles entering and exiting a CTC was examined using a photon with its polarization encoded, as well as a second photon that would act as a past embodiment of the first—or a kind of "stunt double"—in the time-loop simulation they performed. In this way, they could test Deutsch's self-consistency solution by measuring the polarization of the second photon, after it had interacted with the first, averaged across multiple trials. According to co-author Tim Ralph:

> The state we got at our output, the second photon at the simulated exit of the CTC, was the same as that of our input, the first encoded photon at the CTC entrance. This result clearly corroborates the earlier work of David Deutsch. . . . Of course, we're not really sending anything back in time but [the simulation] allows us to study weird evolutions normally not allowed in quantum mechanics.[215]

Even though Ralph and Ringbauer did not actually sending anything back in time, this experiment does contribute to a growing body of work, as well as a growing consensus among physicists, that backward time travel is not only possible but it is also paradox free. Perhaps most notably, these studies demonstrate that no one can change anything in the past, given that their actions already exist as an interwoven part of that past. As such, self-consistency is not a conscious choice made by those who time travel, but instead, is a simple byproduct of existing as part of a different period, which has always possessed embedded elements of another time.

According to Friedman et al., in their 1990 paper *Cauchy Problem in Spacetimes with Closed Timelike Curves*:

> If CTCs are allowed, and if the above vision of theoretical physics' accommodation with them turns out to be more or less correct, then what will this imply about the philosophical notion of free will for humans and other intelligent beings? It certainly will imply that intelligent beings cannot change the past. Such change is incompatible with the principle of self-consistency.[216]

A comical take on this issue of how free will relates to self-consistency and past preservation can be found in the highly cerebral television show *Futurama*. In the episode *Roswell That Ends Well*, the ship's crew accidentally travels back in time to the year 1947, where they find themselves on the Roswell Air Force Base in New Mexico; in an ingenious twist, they become the IFO associated with the famous Roswell crash of 1947. While preparing to leave their ship to look for a device that will help them return to their future time, the lead character, named Fry, tells his distant descendant and nephew, the Professor, that he wants to visit his grandfather, who happens to be stationed at the Roswell Air Force base at that time. The Professor pugnaciously replies to this request by stating:

> Your grandfather?! Stay away from him, you dim-witted monkey! You mustn't interfere with the past. Don't do anything that affects anything. Unless it turns out that you were supposed to do it; in which case, for the love of God, don't not do it![217]

This fictional narrative alludes to a conscious choice by time travelers to maintain self-consistency when returning to the past, as opposed to it being the default condition and an innate byproduct of existing as part of that past. The principle of self-consistency shows that no one is able to actually change anything about the past, but we are physically able to become a part of it and in a way that maintains self-consistency across previously separated slices of block time. In this particular episode of *Futurama*, despite his best efforts at conscious free-will-driven past-preservation, Fry does accidentally end up killing his grandfather, incidentally, while trying to keep him safe in a house that was being used to conduct a nuclear bomb test. However, later, while providing solace to his recently widowed grandmother, he ends up sleeping with her, thus becoming his own grandfather…paradox resolved.[217]

At the risk of sounding like a superfan, this is just one of many clever and insightful scenarios in which *Futurama* probes vexing question related to time, time travel, physics, mathematics, astronomy, and human evolution, which one would not necessarily expect of a cartoon sitcom. However, *Futurama's* creator, David X. Cohen, as well as a number of other writers for the show, hold advanced degrees in physics and mathematics, which include three PhDs, seven master's degrees, and collectively, 50 years spent studying at Harvard University. In fact, one of the show's writers, Ken Keeler, who earned a PhD in applied mathematics from Harvard University, actually proved an unsolved mathematical theorem based on group theory in an episode titled *The Prisoner of Benda*.[218] Though as another writer, Patric M. Verrone once quipped, "We were easily the most overeducated cartoon writers in history, earning critical acclaim, multiple Emmy awards, and a worldwide nerd fan base, but we weren't smart enough to figure out how to avoid cancellation (three times)."[219]

6.6. Physical Pursuit of the Past

The year 2015 marked the 100ᵗʰ anniversary of Albert Einstein's in-fluential papers on general relativity, gravitation, and time.[178 179 180] Over this 100-year period, we have developed an even greater understanding of spacetime, which also includes a number of solutions that would allow for time travel to the past. However, given our primitive position in the grand scheme of time and the long process involved in developing an understanding of the physical mechanisms necessary for creating closed timelike curves, there is still a long way to go. Though with each passing year, decade, century, and millennium, we are certain to unravel ever more mysteries of the fourth dimension. As our collective cultures continue to compound, we will inevitably build upon preceding research and address vexing questions in physics, mathematics, material sciences, engineering, philosophy, and anthropology, which must be confronted as we work to-ward achieving this monumental feat.

Even in the short 100-year period since Einstein put forth a more com-prehensive comprehension of spacetime, subsequent research has spurred an even greater understanding of how one might someday engineer a machine capable of traveling backward through time. Many of these stud-ies have focused on massive bodies, singularities, light, as well as rotating discs and cylinders, which operate under a multitude of conditions. Much knowledge has also developed out of research investigating black holes, as they represent one of few entities in the known universe capable of bend-ing light and time back upon themselves. However, because of the extreme high-energy, gravitational, and electromagnetic spacetime-warping effects of rapidly spinning masses such as these, black holes likely only represent a model upon which we could begin to build a deeper understanding of backward time travel, as opposed to being the mechanism by which we actually do it.

If researchers were to create a singularity, whether it be intentional or unintentional, we are all likely to face certain doom. This would quickly lead to a breakdown of the super-quark composition of the researchers, as well as all other matter that happened to be in and around our solar system at the time. Though in spite of this risk, further research into the physical properties of black holes is warranted. Most notably, because studies investigating how we may send macroscopic objects back through time continue to center on rapidly spinning, electromagnetic, massive and/or high energy, disc and/or cylindrically shaped objects, which often possess time-bending properties similar to that of celestial black holes.

Past and ongoing research would seem to indicate that we have begun to approach an important point in this quest for understanding time and the potential for someday manipulating it in our favor. We draw ever-nearer to the point at which we may finally dislodge ourselves from the seemingly endless unidirectional flow of the fourth dimension, so as to swim jubilantly against the stream, and arrive at a past point along the long and winding river of time. To accomplish this epic undertaking, we would naturally be required to adhere to, and indeed make very good use of, the laws of physics. Most physicists would agree that achieving time travel to the past will also require an all-encompassing master theory, which melds general relativity and quantum field theory. However, when dealing with time travel to the past and the issue of swimming against the raging torrent of increased entropy and the arrow of time, we may also require a sea change in how we conceptualize time itself.

Among other things, we must begin to consider the implications of backward time travel, though not as a violation of causality or any physical laws of the universe, but simply as a practice that runs counter to our normal state of perceived events and our entrenched conceptualization of cause and effect in linear time. Additionally, a metaphysical shift could contribute to our physical pursuit of the past, if we can stop thinking of time as a unidirectional force that is impervious to our desire to transcend it, and stop thinking of the past as something that can be changed, and which therefore requires protection from meddling humans of the future.

It is important to recognize the power of human ingenuity and the increasingly real possibility that we may someday find a way to use this exceptional asset to move at will across the landscape of time. If we do, we would not be changing the past in any way, and these actions should not be considered paradoxical. For if our distant descendants are to ultimately bridge different slices of four-dimensional spacetime, they will simply be uniting both halves of a past-future pair, with each member of their respective time cohort placing omnipresent threads upon the intricately woven web of time.

7

Backward Time Travel

I myself believe that there will one day be time travel; because when we find that something isn't forbidden by the over-arching laws of physics, we usually eventually find a technological way of doing it.[220]

– David Deutsch

Either this is madness or it is Hell. 'It is neither,' calmly replied the voice of the Sphere, 'it is Knowledge; it is Three Dimensions: open your eye once again and try to look steadily.'[221]

– Edwin A. Abbott

7.1. Light Cones, World Lines, Warped Spacetime, and Closed Timelike Curves

Einstein's theory of special relativity demonstrated how space and time are relative to the observer and how high-speed motion alters the rate at which time passes compared to those in different inertial reference frames. Not long after this monumental relativistic revelation, Einstein put forth 10 nonlinear partial differential equations, known as the Einstein Field Equations, as part of his newer general theory of relativity. These equations expanded on the foundation he had laid with special relativity and showed how large masses and energies are capable of eliciting gravitational effects that change the geometry of space and the flow of time around them.

Many solutions to Einstein's original field equations followed and, as stated earlier, a number of these allow for the formation of closed timelike curves (CTCs), which are important with regard to how we may someday come to achieve backward time travel. This is because CTCs create loops that connect different points in time that would, theoretically, allow someone from the future to visit the past. According to mathematician and astrophysicist William B. Bonnor:

> In general relativity a timelike curve in spacetime represents a possible path of a physical object or an observer. Normally such a curve will run from past to future, but in some spacetimes, timelike curves can intersect themselves, giving a loop, or a closed timelike curve (CTC).[222]

Timelike curves represent all possible trajectories of light and matter through time and space, and can be understood as existing within the boundaries of a *light cone*. Because four-dimensional spacetime is difficult to conceptualize and represent graphically, light cones are typically drawn with only two dimensions of space and one dimension of time (figure 7-1). Within this three-dimensional spacetime, it is easier to imagine a single point, beginning at the origin of a beam of light that radiates outward through this geometrical spacetime as an expanding spherical shell. In these two dimensions of space, the light beam propagating outward from *Now* would appear to be an expanding circle. However, if we take cross-sectional slices of the emanating beam at different intervals and stack them on top of one another, they would begin to form a cone, from which light cones derive their name.

Light cones represent all physically allowable events that could occur, or could have occurred, in relation to any other events that lie toward the future or past of the light cone's apex.[223] In this context, we can also conceptualize our own individual *world line*—the path we take through 4-dimensional spacetime—as an historic series of past events, each with a set location and moment in time. This sequence of events can be traced as a long continuous thread that exists within the boundaries of a moving light cone.

We can also abstractly envision all potential paths for our specific world line, since the laws of nature limit these paths to a finite set of possibilities that must exist within the boundaries of our future light cone. More specifically, according to special relativity and quantum entanglement, no mass or signal is capable of traveling faster than light, which means that all matter advancing through spacetime along its own unique world line must lie on or within the moving boundaries of its future light cone.

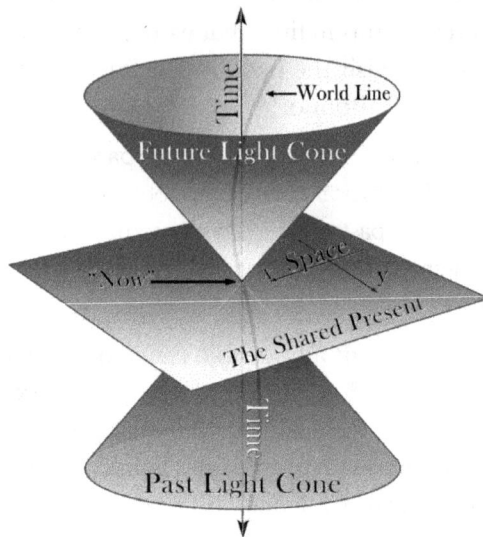

Figure 7-1. *Light cone represented in two dimensions of space (x,y) and one dimension of time (z).*

Light cones can be understood as having the same properties as light, meaning they travel at the speed of light, they propagate like light, and importantly, their trajectory can be warped and curved in the same way as light. This last characteristic of light waves aids astronomers in their search for exoplanets out in deep space, as massive objects warp spacetime to such an extent that we can observe light bending as it passes by them on its way to us. Similarly, when light cones approach an extremely large, dense, or highly energetic mass, their path also becomes curved, and they can begin to tilt (or tip over) while still maintaining their conical shape.

Remarkably, as light cones tip over in association with a warpage of proximate spacetime, world lines within them can deviate from a strictly past-to-future linear trajectory. This means that anyone or anything traveling along a world line that lies within the margins of a tilted light cone in their localized region of spacetime would now be permitted to travel into the global past, by means of this newly formed CTC (figure 7-2). According to the astrophysicists Francisco Lobo and Paulo Crawford:

> A closed timelike curve (CTC) allows time travel, in the sense that an observer which travels on a trajectory in spacetime along this curve, returns to an event which coincides with the departure. The arrow of time leads forward, as measured locally by the observer, but globally he/she may return to an event in the past.[224]

Additionally, according to Leo C. Stein, NASA Einstein fellow in the Department of Astronomy at Cornell University:

> A closed timelike curve (CTC) is a trajectory that's perfectly normal everywhere, always sticking to the rules of moving in a timelike direction, always going (locally) forward in time, and yet ends up back where (and when) it starts. ... The existence of a CTC in some spacetime would mean that a time machine is possible, just by going along that trajectory, and without violating any laws of physics.[225]

7.1.1. Van Stockum Spacetime

Willem Jacob van Stockum was among the first mathematicians to provide a solution to Einstein's field equations that contained CTCs.[226] In a 1938 paper, he described a stationary, symmetrical, cylindrical form of space-time, with a rapidly rotating and infinitely long cylinder of dust surrounded by a vacuum, in which the centrifugal forces are balanced by the gravitational pull of the mass.[224] The rotation of these dust particles cause a frame dragging effect (a.k.a., the Lense-Thirring effect), which is strong enough that light cones tilt in the direction of the rotation.

More specifically, by spinning rapidly along its axis and in association with an intense gravitational field near its center, light cones in and around the spinning mass begin to tip over. As these light cones trace a circular path around the cylinder and draw nearer to its axis of rotation, they angle downward in the direction of rotation, to the extent that the light cone's lower edge now lies below the x-y plane, representing the hypersurface of the present (figure 7-2). Because part of the allowable future contained within the boundaries of these future light cones is now oriented toward the past, time travel to that past becomes possible.[227]

This frame-dragging effect represents a change in the geometry of spacetime resulting from a non-static stationary collection of mass-energy. This means that the rotation of a mass in a stationary field, for example, can cause a change in the curvature of spacetime. Frame-dragging represents yet another example of how Einstein's theory of general relativity has been immensely impactful in shaping our understanding of mass, energy, and gravity, and their effect on spacetime. In fact, it was very shortly after Einstein's 1915 papers on general relativity that, in 1918, Austrian physicists Josef Lense and Hans Thirring derived this important gravitomagnetic frame-dragging effect in the framework of general relativity.[228]

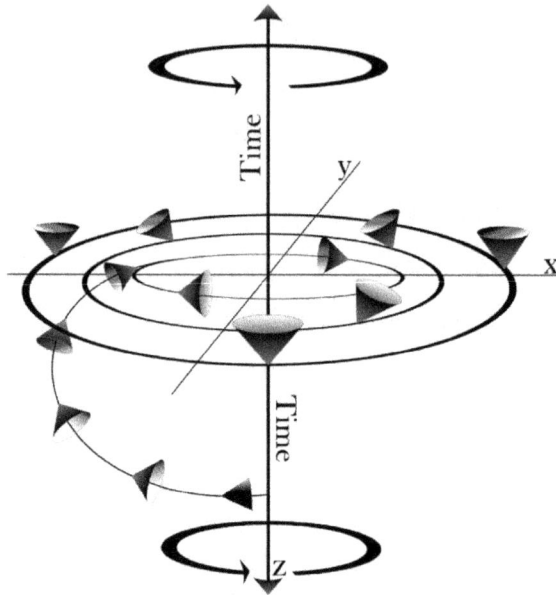

Figure 7-2. *CTC formation in association with warped spacetime, which causes light cones to tip over and world lines within them to become reoriented toward the past.*

Since then—and with ever-evolving technology—we have been able to observe frame-dragging occurring around distant celestial objects with intense gravitational fields such as black holes and neutron stars[229] and even with the rotation of much smaller masses like on our own tiny planet of Earth.[230]

Frame-dragging that occurs in association with Earth's rotation was first demonstrated in 1990 by a team of researchers led by Dr. Ignazio Ciufolini at the National Research Council of Italy and the Aerospace Department of the University of Rome. This represented the first direct detection and measurement of the Lense-Thirring frame-dragging effect, ever since it was initially predicted 100 years ago.[230] Additionally, a more recent study by the same research team, which used millions of laser signals reflected off of two separate satellites above Earth, confirmed 99 percent of the frame-drag predicted for our planet, demonstrating that even the slow spin of a relatively small planet produces a measurable Einstein-Thirring-Lense effect.[231] As described by Dr. Erricos Pavlis of the Joint Center for Earth System Technology, who was a member of the team that made these discoveries:

> General relativity predicts that massive rotating objects should drag space-time around themselves as they rotate. . . . Frame dragging is like what happens if a bowling ball spins in a thick fluid such as molasses. As the ball spins, it pulls the molasses around itself. Anything stuck in the molasses will also move around the ball. Similarly, as the Earth rotates, it pulls space-time in its vicinity around itself.[232]

7.1.2. The Gödel Metric

In 1949, noted logician and mathematician Kurt Friedrich Gödel demonstrated the existence of additional solutions to Einstein's field equations, which also allow for the formation of CTCs.[233] In this paper, Gödel described a basic model of a "rotating universe" with a homogenous matter distribution that rotates at a constant rate around every point within it.[234] Much like van Stockum's Spacetime, this rotation was also associated with light cones tipping over, to the extent that they form CTCs that could allow for backward time travel.

The preconditions associated with this model—involving homogeneity and circular orbits in a rotating universe—make the Gödel metric one of the simplest solutions to Einstein's field equations that allow for CTCs.[238]4However, because it involves a rotating universe with complete homogeny, which is completely unlike our own, Gödel's model—and van Stockum's for similar reasons—are considered to be purely theoretical.[235] As the aforementioned astrophysicist William B. Bonnor points out:

> The first spacetime in which CTCs were noticed was that of Godel.[233] This represents a rotating universe without expansion and requires a negative cosmological constant. As a model of physical reality it can therefore be dismissed because it is unlike the universe we live in. . . . Another simple spacetime containing CTCs is that of van Stockum, which represents a cylinder of rigidly rotating dust; however, the cylinder is of infinite length so it could not be realized in practice.[222]

Even though the van Stockum and Gödel models do not conform to physical reality and cannot directly contribute to the creation of working time machines, they were instrumental in providing an initial understanding of how CTCs can be generated in association with rotating masses. Additionally, these early solutions to Einstein's field equations helped spur a cornucopia of subsequent research into how CTCs may actually be generated, which inches us ever-closer to implementing these collective efforts to achieve time travel to the past.

7.1.3. Tipler Cylinders

It wasn't long after the pioneering work of van Stockum and Gödel that a more realistic model for how these same frame-dragging, CTC-generating effects could be achieved. In a 1974 paper titled *Rotating cylinders and the possibility of global causality violation*, mathematical physicist Frank Jennings Tipler built upon van Stockum's solution to Einstein's field equations. Most notably, Tipler demonstrating how CTCs could be formed using a rapidly rotating cylinder of *finite* length, provided that it was sufficiently large and condensed down to a "ring" singularity:[236]

> ...The gravitational potential of the cylinder's Newtonian analog also diverges at radio infinity, yet this potential is a good approximation near the surface in the middle of a long but finite cylinder, and if we shrink the rotating cylinder down to a "ring" singularity, we end up with the Kerr field, which also has CTL (closed timelike lines). ...
> In short, general relativity suggests that if we construct a sufficiently large rotating cylinder, we create a time machine.[237]

Tipler estimated that the cylinder would need to be roughly equivalent to one solar mass, with a density of 10^{14} g/cm^3, a 10:1 ratio of length to radius, and it would need to be rotating approximately twice per millisecond, with its rim moving at nearly half the speed of light.[238] Tipler cylinders do not require anything to be infinitely large or unrealistically homogenous; however, with this mass, density, and rotational energy—which is roughly equivalent to that of a pulsar or a small black hole—some challenges remain.

For instance, it is thought that the internal gravitation of the cylinder may cause it to become unstable and at risk of collapsing upon itself, and the extreme speed of rotation around its longitudinal axis may cause most forms of matter to explode.[231] So, while they are not purely theoretical, the logistical, physical, material, and engineering challenges may mean that Tipler cylinders do not represent the ideal time machine. In spite of this, Tipler's work represents another important brick laid in a slowly forming foundation upon which we may someday construct a functioning time travel device.

Recent and ongoing work in the field of time travel research is also encouraging. Progressive knowledge is developing around increasingly advanced technology, as well as a better understanding of the material and physical sciences involved in constructing a device capable of reorienting light cones toward the past. Moreover, this modern research continues to raise the foundation initially put in place by Einstein, van Stockum, Gödel, and Tipler, while still centering on the same basic principles of frame dragging caused by rapidly rotating cylinders, rings, spheres and discs. According to astrophysicist William B. Bonnor:

> Since these early discoveries, other spacetimes containing CTCs have been found. Nearly all of these have been regarded as of merely theoretical interest because of some non-physical feature in their composition. Recently, however, there have been published some solutions of Einstein's equations containing CTCs and representing physical situations which in principle could be reproduced in the laboratory or might occur in astrophysics. ...They are not supposed to represent black holes: they could be copper spheres in a laboratory.[222]

Additionally, the astrophysicists Francisco Lobo and Paulo Crawford wrote:

> A great variety of solutions to the Einstein Field Equations (EFEs) containing CTCs exist, but two particularly notorious features seem to stand out. Solutions with a tipping over of the light cones due to a rotation about a cylindrically symmetric axis; and solutions that violate the Energy Conditions of GTR (General Theory of Relativity), which are fundamental in the singularity theorems and theorems of classical black hole thermodynamics[239] ...The tipping over of light cones seem to be a generic feature of some solutions with a rotating cylindrical symmetry.[224]

We are approaching an interesting period along the path of producing a time travel device. Researchers are beginning to move away from investigating the theoretical aspects of generating CTCs, and instead, are focusing on how to apply exact solutions to Einstein's field equations to create these effects in a controlled laboratory setting. After all, if we are to eventually engineer a machine that will allow us to travel backward in time, it will first have to exist in some reproducible physical form. Moreover, if we can look deep into the future and consider the IFOs of our presumed extratempestrial descendants as an indication of where this research may be heading, then all indications are that rapidly spinning highly energetic discs will continue to be an important part of the equation.

7.2. Form Follows Function

A common idiom in biology is that "form follows function," although this is also true of nearly everything that performs a specific function. We can identify morphological elements of an organism and appreciate them in the context of their practical purpose, but we can also understand the form of a machine in the context of its utility as well. For example, a fork pokes food so we can eat it, a chair provides support against the force of gravity so we can sit on it, and a car converts fuel energy into kinetic energy so we can change our location faster and more efficiently than by walking. In the same vein, if extratempestrials are indeed expatriates of time—and IFOs are the devices they use to visit the past—then there must be some *form* elements of these craft that allow for the *function* of backward time travel.

One common characteristic of IFOs, as described by those that have been close enough to observe them, is that they are constantly spinning. This may be a necessary attribute of any airborne object with a saucer-shaped form, given that it would need to rotate at some minimum speed simply to remain stable while hovering or in flight. For instance, if you throw a Frisbee without spinning it, the disc will simply flop around in the air and plummet to the ground. However, if you apply angular momentum to the disc, it will not only fly faster and farther, but it will also remain stable in the air while doing so.

A Levitron, which may actually be a good analogy for the general form and function of IFOs, works by these same basic principles. Patented by U.S. inventor Roy Harrigan in 1983,[240] these spin-stabilized magnetic levitation devices are capable of hovering above a magnetic base with the opposite magnetic pole and can remain stable in this relatively frictionless environment so long as they maintain a minimal speed of rotation (figure 7-3).

The spin of frisbees, levitrons, dreidels, bike tires, etc., is important for maintaining the stability of these object. However, the function of an IFO's spin likely extends beyond simply keeping it stable in the air. As demonstrated above, rapid rotation of dense, massive, or highly energetic cylindrical, spherical, disc, or ring-shaped bodies is a common theme in past and modern research investigating how future light cones could be reoriented toward the past.[224] This indicates that the Frisbee-like form of an IFO may facilitate its function (i.e., spinning at high speed) so that it and its occupants are able to travel backward in time.

Figure 7-3. *Example of a Levitron hovering above its base, by way of spin-stabilized magnetic levitation.*

IFOs are described in close encounters as cylindrical, spherical, triangular, saucer-shaped, and, most commonly, disc-shaped objects. They are also said to rotate around a symmetrical axis and emit high levels of electromagnetic energy. These accounts suggest that rapidly rotating, electromagnetic, disc-shaped bodies are somehow associated with the function of tipping over light cones and traveling backward in time. This supposition may also be supported by examining black holes and how they interact with matter in the universe, given that they are the only known entity capable of causing matter to reverse its trajectory along a linear path through spacetime.

Because most stars rotate and because most collisions have an angular momentum greater than 0, astrophysicists presume that most black holes also rotate at a very high rate of speed.[241] Black holes are also massive disc-shaped objects that absorb and emit high levels of electromagnetic radiation and, most markedly, they are capable of changing the position of matter in space and time.[242] These characteristics of black holes, along with continuing modern physics research of gravitational fields and CTCs generated in association with rotating spheres, solenoids, and most notably, discs,[243 244 245 246 247 248 249 250] indicate that this generalized disc-shaped form is also important for achieving the function of frame dragging (extreme warpage of spacetime) and the capacity to send matter backward through time.

Of course, as always, caution is warranted. In this particular instance, the logical fallacy *post hoc, ergo propter hoc* may apply, which is Latin for "after this, therefore because of this." In other words, even if we view IFOs as evidence that our distant human descendants are destined to someday achieve backward time travel, this does not necessarily mean that we can view our current research as a definitive indication of its inevitable success in creating those same time machines in the future, simply because this contemporary research precedes the development of those future devices. However, even if this current research is not the cause of a device that will allow us to return to the past, these studies still contribute to a deeper understanding of time and time travel, while also helping to advance science and technology as a whole, such as in the development of quantum computers, for example.[251 252 253 254]

If we can appreciate the physical form of IFOs as an indication of their functional capacity to achieve backward time travel, then perhaps even more could be achieved in our quest to traverse time if we took them seriously. If we are in the early phases of developing technology that will ultimately allow us to travel backward through time, then it may benefit us to look forward as well, toward the end result of these current and future efforts. Our present position could epitomize an important component of this self-consistent closed timelike loop in which a future cause (functional time machine) elicits a past effect (knowledge of how to make one) that later results in a time machine that helps spur its own creation, and so it goes.

7.3. Roundtrip Time Travel

Based on our current knowledge of time travel research, it would seem as though we still have a long way to go before we are actually able to achieve backward time travel. Additionally, there is bound to be some time lag between gaining an understanding of the theory involved and actually being able to carry out something as complicated as manipulating matter in spacetime. We would also want to be certain that we had it right, considering that we would be making people disappear from a shared position in spacetime, with the hope that they would later rejoin that same place along the continuum and not be stuck wrestling dinosaurs the rest of their lives.

Implementing any of these theoretical models on a grand scale will first require an enormous amount of precise mathematical calculations, detailed engineering, and, most likely, a lot of materials and technology that don't yet exist. Though as stated previously and as echoed by David Deutsch in the quote that begins this chapter, humanity has consistently demonstrated that if something is not explicitly prohibited by the laws of physics, then we are likely to eventually find some technological way of doing it. This is apparent from the increasingly complex technologies that we have developed since crafting the first stone tools nearly 3.3 million years ago, as well as the seemingly unimaginable feats that our species has achieved ever since, particularly during the recent human past.

For example, on November 12, 2014, a massive team of scientists was able to safely land an advanced spacecraft on a comet hurtling through our solar system, which is an accomplishment that would have seemed impossible even 100 years ago. It is astounding just how fast we advance our scientific understanding and ability to engineer highly sophisticated technological devices. Looking at recent human history, it is also easy to conceptualize how we someday develop the knowledge and machinery necessary to accomplish backward time travel, particularly if this accelerating rate of technological development continues into the future.

There is undoubtedly a great deal of math and engineering involved in something as challenging as intercepting a comet whizzing through space. However, this process still takes place within our shared position in time, which allows for the precise calculation of where in space each of these things are and will be in relation to one another at very precise times. By contrast, a device sent backward in time would inevitably disappear from the reference frame of those who were involved in sending it back. As a result, we could potentially lose all ability to analyze what happened to it or to communicate with anyone brave enough to go with it, given that the time machine and pilot would now be moving in the opposite direction of everyone in their home time.

The veil of time may well add an extra element of uncertainty and logistical difficulty to both testing and actually carrying out roundtrip time travel. For instance, how do we gauge just how much of which parameters are required to accurately send someone to the intended time period? How do we measure the effectiveness of this time placement, particularly if the device itself disappears from our own position in spacetime? What happens to the device and the data associated with its journey back through time? With this in mind, early tests may require that we first have a firm grasp on how to send then return a completely autonomous device to its time of departure. This way, the results of roundtrip voyages could be thoroughly analyzed, long in advance of any attempt at sending humans back through time.

7.3.1. Time Dilation, Closed Timelike Curves, and the Return Trip Home

Some physicists and logicians have argued that it is not possible to visit a time period that predates the development of the time machine used to travel to it, since the device that allows this manipulation of spacetime does not yet exist. Others believe that travel beyond this point in the past is possible, but only back to the time at which the very first time machine was developed. However, it is difficult to imagine how a time machine would simply begin to disintegrate into all its original disarticulated parts as it approached the time at which it or any other time machine was first created, particularly considering that the time travel device—as well as everyone and everything in and around it—had already become dislodged from global spacetime within their localized reference frame.

While traveling backward through time, a time travel device becomes separated from the sustained forward trajectory of everyone else. Because of this temporal exodus, how far it goes while traveling backward should not be limited by any individual event that exists outside its localized reference frame, including the creation of the time machine or the conception and birth of anyone inside. Additionally, through the mechanism of time dilation—resulting from high-speed motion and/or increased gravitation—it may be possible to speed the rate at which a time machine moves backward through time, potentially allowing it to go even farther into the past at a faster rate.

Einstein's theory of special relativity demonstrated how high-speed motion, relative to those in a rest frame, allows for time travel to the future. But, if a time machine moves at a high rate of speed while light cones are oriented toward the past—as measured locally by the observer—this would be expected to move the craft even deeper into that past and perhaps far beyond the origins of both themselves and their time machine. In other words, if a rapidly rotating disc with the right mass-energy-rotation parameters were to facilitate backward time travel, then sustained high-speed motion while spinning and while light cones are tilted toward the past may allow it to go even deeper into the future of that past.

Einstein's theory of general relativity later demonstrated how this same time dilation effect can be achieved in high gravity. So positioning the time machine in a region of space with a sizeable natural or manmade gravitational field might also allow for deeper penetration into the distant past. Additionally, traveling at high speed or experiencing high gravity for some duration may be important for helping a time traveler return home after visiting the past. In this way, time dilation, resulting from high-speed motion and/or high gravity, would allow a time traveler to speed back through time to their original present, which may make getting back to the future a far easier pursuit than going backward through time.

7.4. Attend to Your Configuration

A fascinating aspect of reports of close encounters is that IFOs are commonly observed appearing or disappearing, often while maintaining the same position in the sky. This is important in the context of the current time travel model, because if something abruptly appears or disappears in our three observable dimensions of space, it is a very good indication that it has just changed its position in the only other observable dimension —time. This regularly reported occurrence, while seemingly bizarre from the standpoint of our conventional notion of linear time, is actually an expected outcome of time machines entering or exiting a specific region of spacetime.

A superb heuristic device for understanding divisions among the three dimensions of space and one dimension of time—and how the passage through and between these is observed differently among members of each—can be found in the multifaceted book *Flatland*.[221] Written as mathematical satire in 1884 by British liberal theologian Edwin Abbott, *Flatland* provides an insightful look into how perception shapes our sense of reality, regardless of whether it is an accurate interpretation of the world around us.

The second quote that begins this chapter was taken from a specific section of Abbott's book, in which a two-dimensional being living in Flatland observes a three-dimensional sphere passing through his more limited plane of existence. Initially, the two-dimensional entity perceives the three-dimensional sphere as only a small dot that suddenly appears before him. However, the dot soon begins to grow into an oddly expanding circle. Upon reaching its maximum diameter, the circle then begins to contract, until it returns to its original dot-like form, just before disappearing entirely from view.[221]

Although the two-dimensional being is capable of viewing this three-dimensional traveler in some rudimentary capacity, the 2-D being is unable to see the 3-D traveler in his entirety or to understand how the 3-D traveler may perceive his own movement through the less complex 2-D world of Flatland. This fictional scenario is somewhat analogous to how we, as three-dimensional creatures embarking on a rectilinear path through time, may observe 4-D travelers entering, exiting, or migrating through our specific region of spacetime. The abrupt appearance or disappearance of an object is currently outside our realm of common understanding, as we continue to move collectively forward, along with everything else, unable to change our position in time. In this sense, pre-time travel humans may be somewhat akin to the two-dimensional inhabitants of Flatland, left to ruminate on the oddities associated with observing more complex beings who have developed the capacity to move breezily through the fourth dimension.

Reports of IFOs appearing or disappearing in the sky are quite common. In fact, a simple search for the words *appear* and *disappear* among vetted reports filed by the National UFO Reporting Center (NUFORC) over the previous year, returns over 290 descriptions of this occurrence.[255] Naturally, many of these tallied instances were found in the search simply because people commonly use the word "appear" to describe something they see all of a sudden, regardless of whether it appeared to have appeared out of nowhere. To account for this potential source of content analysis bias, these 290 reports were examined individually to assess the intended meaning behind the usage of these words (appear and disappear). Though even after this corrective procedure, 229 reports remained that described a light source or actual craft that seemingly appeared or disappeared into or out of thin air (see table 7-1 for select examples).

As stated above, the rapid appearance and disappearance of these objects and lights likely represent our extratempestrial descendants, moving in and out of the fourth-dimensional reference frame of these befuddled bystanders. This behavior, as well as the physical properties of the IFOs while appearing and disappearing, is exactly what one would expect of a device that is dipping in and out of different regions of block time. In fact, Stephen Hawking, a long-time opponent of backward time travel, even acknowledged this as a widely held view of time machines, despite his reassertion that time travel will never actually happen in this way or any other. In part one of a six-part reality-show-themed PBS miniseries called *Genius*, Hawking is quoted as saying:

Most people think of time travel as a sudden disappearance from one moment in time and an instantaneous arrival in another, usually involving a cleverly engineered time machine. I hate to be a killjoy, but I doubt it will ever be possible to jump through spacetime in this way. [256]

The multitude of reports of spherical, cylindrical, and disc-shaped craft—with physical properties ideally suited for time travel—which are commonly observed suddenly disappearing or instantaneously arriving in our own time, may represent evidence to the contrary. In fact, these epitomize a cleverly engineered time machine, presumably developed over countless millennia by our even cleverer descendants, which are now observed to appear and disappear in the presence of us, their distant ancestors. Furthermore, many reports describe different colors of light emanating from the IFOs as they emerge and vanish in the skies above. These light emissions and color spectrum changes could also be a testament to the time travel capabilities of these craft, while also speaking to the vital role that electromagnetism plays in helping to facilitate this incredible function.

7.5. Electromagnetism

Changes in the visible light spectrum, as IFOs are seen appearing and disappearing in the sky, could be related to changes in their speed of rotation while entering or exiting our region of spacetime. More specifically, one would expect the visual light spectrum around an IFO to become blue-shifted, meaning a displacement toward shorter wavelengths, as it rotates faster and more heat is generated in association with increased atmospheric friction. As an IFO slows its speed of rotation and generates less heat from friction with air particles in the atmosphere, a redshift would be expected to occur, meaning longer wavelengths toward the red end of the visible light spectrum. In this way, changes to the light characteristics of IFOs may reveal something about their status as arriving in or departing from our time.

In addition to these visible light characteristics, high electromagnetic fields are also frequently detected in and around IFOs and places they have been. For instance, during cases of close encounters, it is common for cars and car radios to initially become increasingly erratic and then completely shut down as an IFO approaches. This phenomenon was detailed in the aforementioned abduction case of Betty and Barney Hill, who also noticed that a compass spun wildly when placed near small

Reported IFO Sightings

Date/Time	City	State	Shape	Duration	Summary
11/7/14 20:35	Wilmington	NC	Light	10 minutes	Two stationary orange lights in the sky flickering on and off, fading to a reddish color before disappearing.
8/28/14 22:00	Seattle (lower Queen Ann area)	WA	Circle	2 minutes	6 yellowish glowing orbs in night sky fly strange patterns over Seattle, then perform a synchronized disappearance. I came upon 5 gentlemen on first ave in Seattle looking at something, they claimed they were UFOs. I watched the 5-6 spherical lights move up and down, left and right, into each other (seemingly) or over each other and then separate again. What was most interesting was the synchronized disappearance. In a jagged line, they vanished one by one. The one on the left moved down slightly for less than a second, then disappeared in mid-air. The one just to the right of that one did the same. And so on and so forth in a pattern until all were gone…
8/25/14 22:09	Escondido	CA	Sphere	3-4 minutes	Orange/red pulsating sphere suddenly appeared in the sky and continued dimming and brightening before fading entirely. Never moved.
8/10/14 22:00	New York City	NY	Disk	1 minute	Oscillating black-ish/gray spherical object hovering in the sky and disappeared.
7/29/14 02:00	Lumberton	NC	Light	2 hours	Bright white light that would change colors, disappear, & reappear over S.E. N. Carolina witnessed by 4 people over 2 hrs.
7/21/14 23:10	Oshawa (Canada)	ON	Circle	10 seconds	Craft appeared to be silent, four then five rotating red lights, moving west to east for 10 seconds, changing to orange then disappearing.
7/9/14 21:30	Johnstown	PA	Circle	30 seconds	Saw three large red orbs moving slowly above Johnstown that simply disappeared after changed colors a few times.
7/4/14 22:00	Bartlett	IL	Egg	7 minutes	Bright green/red HOVERING object 5 mins west sky (during fireworks to north) DARTED with impossible speed SE, hovered, disappeared.
6/28/14 23:30	Rainelle	WV	Light	4 minutes	Saw 2 large yellow lights, with red light strip on bottom, traveling NE, stopping, hovering, separating and then suddenly disappearing.
6/27/14 23:00	Columbus	OH	Disk	10 seconds	Floating disk with red, yellow, green, and white lights appeared and then vanished with a small white flash but NO noise.

	Reported IFO Sightings, continued…				
Date/Time	City	State	Shape	Duration	Summary
6/19/14 21:00	Wayne	NJ	Circle	13 minutes	2 glowing circular crafts high in sky, not satellite or plane by deductive logic, disappeared in a flash.
6/1/14 21:27	Greensboro	NC	Unknown	~2-3 minutes	Taking daughter for a walk and saw two reddish orange lights spread apart bigger than a football field then disappeared into thin air!!!
5/29/14 23:35	Zion	IL	Light	10 seconds	Large bright light appeared for 10 seconds and then disappeared. (NUFORC Note: Witness formerly in U. S. Navy for 20 years. PD)
5/14/14 20:00	Bend	OR	Disk	20 seconds	Reflective, oval, disk-shaped object with a haze around it, appearing and disappearing, travelling at an astonishing rate of speed.
5/6/14 20:35	Gadsden	AL	Cylinder	10 minutes	Strange cylinder-shaped object appeared huge in size and illuminating white lights over hwy 278 in Gadsden; disappeared in thin air.
4/28/14 23:30	Sunrise	FL	Sphere	30 seconds	Sphere in the sky over Sunrise, FL, emitted lights that constantly shifted and changed colors before disappearing.
4/17/14 14:00	Columbus	OH	Sphere	5 minutes	Silver object that jumped across the sky so many feet. It moved very quickly would disappear and reappear. Then it's gone.
3/9/14 22:00	Canton	MI	Sphere	2 minutes	3 orange orbs 1,000 feet off ground 100 yards apart in triangle formation then disappear.
1/6/14 19:00	Georgetown	KY	Circle	30 seconds	Two orange lights moving due south to due north combining as one then disappearing with no sound.
1/1/14 00:53	Reseda	CA	Circle	<1 minute	Three orange lights on New Year's Eve form triangle, align themselves then disappear.

Table 7-1. *IFO sightings reported to NUFORC between 1-1-2014 and 1-1-2015. Showing 20 of the 229 total sightings involving the observed appearance/disappearance of an IFO, which may indicate their passage into, out of, or through the observer's localized region of four-dimensional spacetime.*

circles on their car following the abduction event. The ubiquity of these reports indicates that the electromagnetic force may be instrumental in the operation of IFOs, for their ability to bend spacetime, and as a key component of their propulsion system. In fact, there is already some indication that electromagnetism could be a valuable tool in our pervasive fight against gravity.

Toward the beginning of this century, a mathematician and aerospace engineer named Dr. Ronald Evans took the reins of Project Greenglow. Backed by BAE Systems—one of the largest defense contractors in the world—Evans' role as head of this project was to research ways to counter the force of gravity using electromagnetism, in order to engineer more advanced aircraft. The underlying notion was that electromagnetic control of gravity could lead to the development of a new method of propulsion, which would launch a subsequent revolution in the commercial, public, and private aerospace industries.[257]

Although this research did not directly result in a practical form of aeronautic gravity control (that we know of), there are a number of indications that this could be a fruitful and highly lucrative pursuit. Evidence for this comes from the fact that there is an immense difference in the strength of the electromagnetic force relative to that of gravity. In fact, according to Dr. Clifford Johnson, a professor in the Department of Physics and Astronomy at the University of Southern California:

> We tend to think of gravity as very strong—after all it's what binds us to the earth. But actually of all the forces we know in nature, gravity's the weakest. . . . Let me give you a number. It's 10 to the power 40 times weaker than electromagnetism, that's a one with 40 zeros after it![258]

The extreme difference between the force of gravity and the electromagnetic force points to the role that the latter may someday play in helping us finally break free of the antiquated, bernoullian, vector-based, combustion engine propulsion technologies of the present era. If we are able to develop an aircraft that can harness and control the electromagnetic force against the weaker force of gravity, this could fundamentally transform the way we travel and the speed with which we get around the world as well. Furthermore, one of the best shapes for opposing Earth's gravity—once we no longer require wings and tubes to fly—is likely to be a disc shape, which can counter the force of gravity more homogenously across its symmetrical base, while rotating in the air to maintain lateral stability.

As stated above, the strong electromagnetic force may be critical in helping us design and construct a craft capable of bending and warping spacetime to allow for time travel to the past. Because electromagnetism is 10,000,000,000,000,000,000,000,000,000,000,000,000,000 times stronger than gravity, it could be a key component of forming CTCs through frame dragging and rotational energy. Manipulating the electromagnetic force could help to bend spacetime enough so that light cones are reoriented toward the past, without having to maneuver masses equivalent to that of black holes, which may otherwise be needed to achieve the same level of spacetime warpage using the force of gravity.

If IFOs provide a glimpse into the future state of human technology, then perhaps it will ultimately be shown that BAE, Dr. Ron Evans, and Yevgeny Podkletnov—a Russian engineer who first detected antigravity over the surface of a spinning superconducting disc in the late 1990s[259] —were all on the right path all along. The form of IFOs would seem to suggest that our species will eventually harness the awesome power of the electromagnetic force on a grand scale for the function of advancing aeronautic propulsion and, presumably much later on, for the purposes of traveling backward through time.

7.6. It Is Knowledge; It Is Four Dimensions

These and numerous other examples of recent and ongoing research involving electromagnetism, levitation, propulsion, CTCs, etc., indicate that our rudimentary understanding of time, and the engineering requirements associated with making a craft that can travel backward through it, are perhaps further along than it may seem. However, even if we do master time travel in the relative near term, we are not likely to see early versions of this technology anytime soon. From a time-travel standpoint, there is little foreseeable benefit to visiting a period so close to one's own time, considering that we are likely to retain accurate records of events between now and then, which would make any trip to the recent past somewhat redundant, at least from an information gathering standpoint.

A novel high-tech aeronautic device such as this is also likely to be so highly coveted by private industry and other members of the military industrial complex that this technology would undoubtedly be kept under very tight security. In fact, right now, BAE systems could be testing advanced aircraft with electromagnetic propulsion systems stemming from Evans' time at Project Greenglow. However, because of the proprietary value of this intellectual property and the tremendous amount of money that could be made from being the first to implement it, no one would ever know until they wanted them to.

This issue of technological secrecy, in both an intra-temporal and inter-temporal context, raises a number of important questions. For instance, how long would sophisticated technologies that allow for better propulsion systems and/or the ability to time travel need to be hidden from the general populous, if this is to eventually become a part of our shared reality? If IFOs are to someday exist—whether they have been equipped for time travel or are simply used as a better alternative to airplanes—then there is an inevitable point in the future when knowledge of them will be widespread and eventually, entirely normalized.

It is interesting to speculate about how this "new" information might be disclosed, particularly if these technologies had already been in covert development for some time prior. For instance, an abrupt discharge of information would suddenly challenge the worldview of those who had grown accustomed to incremental technological advancement, without any major periods of punctuated equilibrium. By contrast, a gradual release of information may be more palpable for the general population, though this might not always be an option, especially if whoever possessed it was getting beat in an international conflict, for example.

Regardless of how it is developed and disseminated, after these advanced aeronautic technologies are introduced and become normalized within global society, there may no longer need to be any form of secrecy surrounding them. Broad dissemination of this information could potentially emancipate our distant descendants and allow them to move more freely within and across any time period. For example, if someone from the year 150,000 AD were interested in going back to observe a notable event that took place in the year 130,976 AD, they may not actually have to sneak around at all, if their technology and existence as a member of the same species from a future period were already broadly understood as such.

This scenario, in which more ostentatious instances of intertemporal interaction are initiated and sustained, would be expected to usher in a new phase of human history, one in which we begin to integrate ensuing stages of our biological and cultural evolutionary future with those of the past. This would undoubtedly add complexity to our currently linear biocultural evolutionary line, as future people and things could influence the development of those same people and things, when introduced into a past that they have always been a part of. Integration among past and future periods would inevitably force us to reconsider time and transition it away from being a misunderstood, uncontrollable, ambiguous entity, to one involving a more comprehensive conceptualization of intertwined events, structured around a more cyclic system of cause and effect. If this occurs, we are likely to begin to see the world more like the inhabitants of Flatland: opening our eyes to look steadily upon a more transcendent spacetime, where we have always been the interdimensional beings of our own perpetual reality, appearing and disappearing at various points throughout a deep and multifaceted hominin evolutionary past.

8

Bipedalism and Biocultural Evolution

The right question to ask from a Darwinian perspective is what was it about bipedalism that was so advantageous? Why did that adaptation ultimately lead to a species, Homo sapiens, that has come to dominate the planet today?[260]

– Donald Johanson

8.1. A Cautionary Tale of Teleology and Time

The paradigm put forth in this text focuses on the notion that the quintessential alien archetype is simply a future form of ourselves, returning to study us as part of their own ancestral hominin past. A deeper appreciation of this model remains veiled by our current position in time and our present inability to travel to different periods of it. We may be able to garner some details of our future descendants from ephemeral glimpses of them, as they occasionally descend upon us from that future. Though the best starting point for visualizing how this intertemporal interaction may come to pass is to first look back through our own past, examining persistent patterns of cultural and biological change across the last six million years of hominin evolution.

Making predictions about the future state of humanity is often dismissed as *teleology* among evolutionary biologists. This is the idea that purpose and design exist in the natural world; i.e., there are immanent ends associated with long-term processes and the changing state of organisms through time. However, this criticism is not applicable to the current model and is somewhat detrimental to the pursuit of knowledge more broadly. This teleological critique overlooks how the same conspicuous trends, which epitomize hominin evolution over the last 6–8 million years, are very likely to continue into the future, regardless of how they are perceived or interpreted by us at any point along the way.

The aim of this discussion is not to explain the big-headed, big-eyed, small-faced characteristics of time traveling humans in the context of what purpose these traits may serve or how they exemplify some kind of ideal state of being. To the contrary, a number of detrimental conditions have arisen in association with these and other longstanding anatomical and cultural trends specific to the hominin lineage, including those described earlier in reference to the article *The Perils of Being Bipedal.*[261] Rather, the focus of this discourse is to examine these tenacious trends—which have persisted throughout hominin evolution in spite of evolutionary tradeoffs and an ebb and flow of social and environmental conditions—with the goal of demonstrating how these evolutionary patterns are likely to persist, and continue to accelerate into the distant future.

It is also rather inequitable that certain attempts at predicting future conditions are acceptable, while others are not. For instance, we are free to discuss and make predictions about the future of finance, economics, politics, meteorology, climatology, and even the evolutionary trajectory of other organisms. However, it is taboo when predictions are made regarding the human future, regardless of how enduring past trends may be. Meteorologists make daily predictions based on past trends and recurring climatic patterns, and even though they are often wrong about near-term weather phenomena—and at a far higher rate than would be acceptable in any other field—it remains a widely accepted form of scientific prophecy.

Each year virologists speculate about and attempt to predict the forthcoming allele frequency of various strains of the influenza virus in an effort to make the most effective flu vaccine to combat it, though they often fail in their predictions as well. For instance, according to the most recent estimates from the Center for Disease Control and Prevention, effectiveness of the influenza A H3N2 vaccine during the 2014–2015 season was an abysmal 18%.[262 263] As a member of the very large group of people who received the flu shot that year but still contracted the flu, this was a painful reminder of how difficult it is to predict the evolutionary trajectory of any biological entity, particularly for those with very high mutation rates (i.e., high frequency of coding errors during gene replication).

Looking into the darkness of the future with only a flashlight from the past is problematic for any field, since we are currently only able to see what has already transpired, as we remain shackled to a linearly oriented arrow of time. Given this across-the-board limitation, all industries and academic fields should be treated equally in their efforts to predict things that only the future can fully reveal. Furthermore, teleological censure should not be invoked if the past and future exist as highly integrated entities. These networks of self-sustaining cycles, spanning small segments of a far greater spacetime that—because of their perpetual interaction and contorted connection between cause and effect—exist outside the realm of premeditated purpose and design.

If the IFO phenomenon does involve an interconnectedness of people across different periods of time, then making predictions about where we are going based on where we have been is only half of the equation. Because this proposed model centers on intertemporal interconnectivity and the possibility of observing the technology and people of the distant future, then we may be provided a unique opportunity to simply connect the dots, as opposed to shooting wildly into the wind with regard to where we think we might be headed. For if we were to candidly consider the possibility of intertwined time, then the opacity of the future may begin to clear among those for whom it is already the much more lucid past.

8.2. Bipedalism, Craniofacial Evolution, and Cerebral and Cognitive Convergence

A common mantra in investing—and an SEC-mandated disclaimer no less—is that "past performance does not necessarily predict future results."[264] Unforeseen events can cause major disruptions to the stock market and send an individual stock, sector, or the entire system plunging into a state of chaos on a whim. Major events can also affect how organisms evolve, such as asteroid impacts, geologic events, or even the ascent of *Homo sapiens* in the Anthropocene, which had, and continues to have, a major effect on most lifeforms on this planet, including ourselves. Though comparatively, past trends are a better predictor of future results in biology, simply because evolutionary changes are always built upon prior genotypic and phenotypic variants.

Even with this more stable process of aggregating evolution, there remains the possibility that future environments could suddenly change, ushering in a new set of selective pressures. However, changing environments have not caused a broad-based reversal of the two most prominent craniofacial trends in human evolution: encephalization (increased brain expansion) and facial orthognathism (reduction and retraction of the mid and lower face). Our hominin ancestors migrated to new environments, watched them change, and modified them to suit our own species-specific needs more recently. Though in spite of vast changes to our environment and to our social, economic, and political systems, encephalization and facial orthognathism have persisted and have even accelerated throughout hominin evolution.

A great deal has changed with regard to our physical, cognitive, and behavioral characteristics since we last shared a common ancestor with our closest living cousins, the common chimpanzee (*Pan troglodytes*), and the pygmy chimpanzee (*Pan paniscus*). This common ancestor of modern chimps and humans did not look like a chimp or a human, though it certainly looked a lot more like them than us now. Even though chimpanzees have evolved genetically, morphologically, cognitively, and behaviorally over the last 6–8 million years, they have undergone far fewer changes when compared with our own long and complex evolutionary history. This is because chimpanzees remained in the rainforest environments of Western and Central Africa and therefore did not encounter additional selective pressures from the variable and changing environments that hominins encountered in the eastern and southern parts of the continent of Africa.

Because chimpanzees evolved in a relatively stable environment, they retained more *ancestral traits*, meaning characteristics that are indicative of our shared common ancestor. Conversely, and in response to different and changing environments, our hominin ancestors began to develop more *derived traits*. This term refers to characteristics that are unique to our own evolutionary branch, including such things as a fully opposable thumb, reduced body hair, larger brains, and reduced facial projection, among others.

A number of factors have contributed to observable differences between humans and chimpanzees since we bifurcated from the same budding branch on the tree of life, though the most influential of these was undoubtedly our shift toward habitual upright walking. As previously mentioned, bipedalism is what defines our lineage relative to that of chimpanzees and our other primate relatives. Standing and walking upright also initiated a number of other important morphological and behavioral changes that primed us for success as we began to make and use tools and later migrate out of Africa into nearly all other parts of the accessible world.

One of the earliest morphological changes associated with bipedalism was a shortening and broadening of the pelvis, which helped us maintain balance while also creating a basin-shaped receptacle to hold our now-vertically oriented guts and babies. Later changes in hominin postcranial morphology (i.e., our skeletal anatomy from the neck down) included a heel and arch adapted for two-legged walking; a lengthening of the lower limbs that gave us a longer and more efficient gait; an inwardly angled femur that put our center of gravity more toward the center of our bodies; a big toe that was more in line with our other toes; and an S-shaped spinal column, which helped with balance and shock absorption while standing, walking, and running bipedally.

Indications of bipedalism can also be seen in the skulls of our early ancestors, as this shift toward habitual upright walking resulted in a number of subsequent craniofacial changes, including a larger, more complex, and more intelligent brain. As mentioned previously, one important change was that our faces began to rotate downward. This helped to realign our visual plane with the horizon so we could see where we were going. A consequence of this downward and backward rotation of the viscerocranium (mid and lower face) was that our eyes and faces eventually became situated directly below the frontal lobes of the brain in modern humans, which is an incredibly rare craniofacial configuration among mammals. Along with this downward and backward rotation of the face, the foramen magnum—again, the large hole at the base of the skull through which the spinal cord passes—also began to move anteriorly; i.e., toward the middle of the base of the skull. This broad-based reorientation of the visual plane and associated forward movement of the foramen magnum were a vital part of how our brains came to be so large in relation to other animals on this planet.

The extreme bending and flexing of the skull that occurred in association with our shift toward habitual bipedalism is known as *basicranial flexing*. Described earlier in the context of what happens to a slinky when the two ends are bent down and toward each other, basicranial flexing helped open up new space in the upper skull that could then be occupied by an increasingly larger brain. More specifically, as the face retreats downward and backward (left side of the slinky in this hypothetical), and as the foramen magnum moves downward and forward (right side of slinky), more space opens up toward the top of the skull, as it bows upward and outward in association with a flexing of the entire cranial base below (figure 8-1).

Taken together, our large and highly sophisticated brains were made possible by, or are at least intricately linked to, something as simple as our early hominin ancestors beginning to stand and walk upright over 6 million years ago. In this way, bipedalism and basicranial flexing were

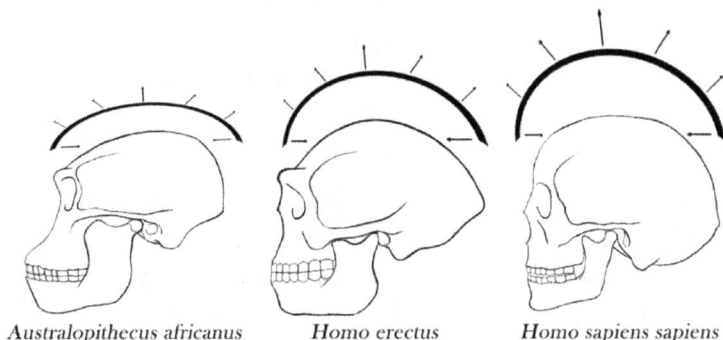

Australopithecus africanus   Homo erectus   Homo sapiens sapiens

Figure 8-1. *Effects of basicranial flexing on size and shape of the neurocranium in hominin evolution.*

important early contributors to increased brain size and more advanced cognitive function in human evolution. Over time, additional factors contributed to an even greater upsurge in hominin cranial capacity and further reorganization of our brains, which correlates with increased intelligence, as indicated by the progressively more complex tools, technologies, and societies of succeeding hominin species with larger brains. By all accounts, this accelerating rate of brain expansion/reorganization and increased technological sophistication, which itself helped contribute to cerebral evolution and increased cognition, are likely to continue long into the human future.

During early hominin evolution, the skeletal potential for larger brains began to develop in association with bipedalism and basicranial flexing. However, we were still lacking a number of important external conditions that contribute to differential rates of selective fitness among individuals and, over time, increased intelligence in the population as a whole. Such things as rapidly changing environments, long-distance migration, changes in group size and social complexity, language, and the important role of culture, tools, and technology, as they relate to each of these other factors in the context of natural and sexual selection.

As previously stated, long-distance migration into new and different environments correlated with an increase in hominin brain size and the development of increasingly sophisticated tools and lifeways. Certainly, correlation does not equal causality. However, it may be safe to assume that these new challenges elicited some causal change with regard to our ability to adapt using new and smarter solutions. Additionally, many of these solutions (e.g., clothing, fire, tool types), appear to be in direct response to new climates, predators, food resources, and other aspects of the various environments we migrated to throughout the old world. It was this intense and rapidly changing set of circumstances and selective pressures that likely united our innate ability to survive and reproduce, with the increasingly important ability to adapt to new and changing environments using our rapidly evolving cognition and culture.

We continued to adapt biologically throughout this time, though innovation and cultural change were beginning to play a larger role in helping us adapt to, migrate between, and survive in a vast array of diverse environments. This cultural plasticity began a new stage of natural selection in humans, which favored those who could think their way out of problems through the development of better clothing, tools, hunting strategies, political and economic alliances, communication, and a general sense of enlightenment relative to those around them. This ability to outthink their immediate social and physical environment would confer a greater probability of survival and, importantly, reproductive success. As those who excelled in these capacities would be expected to garner more mates and would likely be better at keeping their offspring alive, thus increasing their selective fitness as well as the frequency of their judicious genes over subsequent generations.

It may seem like circular reasoning to say that bigger brains developed out of cultural innovation, when innovation itself is the product of a larger, more complex, and better-organized brain. However, natural selection works on *individuals*, and not populations. Therefore, if certain individuals possessed a heritable brain structure that allowed them to invent better tools, develop more effective hunting and warfare tactics, better gathering techniques, child-rearing practices, or better social skills in politics, economics, and mate attraction, then whatever aspect of their brain that contributed to those beneficial characteristics would be expected to increase in frequency throughout the larger population over time.

It is individuals that solve problems, develop solutions, and drive innovation. The ones who are most successful in these endeavors are often rewarded with increased reproductive opportunities and, more recently, increased material wealth relative to other members of society. These individual contributions add to the collective knowledge of the population, which, over time, raises the entire competitive playing field among concurrent conspecifics, thus imposing further generation-specific requirements upon those born into each progressive stage of human evolution.

This compounding cultural complexity is analogous to an exponential function plotting population growth, radioactive decay, or the growth trajectory of a fixed interest rate savings account, for example. At first, interest will accrue slowly, because of the small amount of money from which the interest is drawn. Though over time, the account grows larger with increased rapidity, as the same rate of interest returns larger sums, drawn from increasingly larger principal amounts. In this context of intra and inter-cultural competition as a driver for human innovation, intellect, and encephalization, more people equates to more individuals that can contribute to the collective pool of knowledge for the entire group. Though as the innovations of these innovators begin to spread and become normalized, the playing field is once again leveled, albeit at a slightly higher level and with a novel set of competitive requirements relative to those that came before.

The onus is on the individual at each stage of this unremitting battleground of ideas and implementation to out-compete other members of that same stage of sociocultural progress. With increasingly complex societies and technologies, our ancestors continued to compete for the same resources and mates, but with a different set of physical, cognitive, and learned advantages over each other. It is this aspect of compounding competition, in association with cranial flexing and other factors,[265] which helped advance the accelerating rate of encephalization and increased cerebral and cognitive complexity throughout hominin evolution.

Again, the robust correlation between increased brain size and cultural complexity does not inevitably indicate causality, or reciprocal causality for that matter. Though the temporal association between these variables certainly raises questions about what contributes most to the runaway brain in long-term hominin evolution. Changing natural environments, human-induced social and physical environments, competition within and among societies, as well as changes to our life-history cycle and pattern of growth and development, have all contributed to our expanded cognition and unique form of biocultural complexity. According to Hublin et al. (2015), in an article titled *Brain ontogeny and life history in Pleistocene hominins*:

> Human-accelerated sociocultural evolution can be seen as niche construction, in which our large brain plays a central role. Its unparalleled increase depended on the establishment of peculiar developmental and demographic patterns, including protracted growth and delayed reproductive life. As human brain maturation extends over a very long period of time and is shaped by the interactions

of the individuals with their environment, the uniquely long human
developmental pattern has reciprocally contributed to the increase
in our behavioural complexity.[266]

Once the runaway brain train got rolling down the track, it began a
type of brain-behavior feedback loop. Our larger and more complex cerebra contributed to more advanced behavior, while this more advanced
behavior subsequently contributed to a larger and more complex brain.
Admittedly, this self-licking ice cream cone explanation for the synergistic
relationship between compounding cultural knowledge and cerebral expansion may seem like a post hoc ergo propter hoc non sequitur. However,
both the fossil and archaeological records demonstrate a strong association
between these coalescing cultural and cerebral characteristics, as well as an
accelerating rate of change in each, and particularly during more recent
stages of human evolution.

As stated above, one might also expect these long-term trends of increased brain size and greater cognitive complexity to continue into the
future, considering that the rate of culture change continues to accelerate
locally, regionally, and globally. Additionally, these enduring trends may
represent another facet of the human condition that links our distant
past to our distant future. Our unique form of biocultural evolution—in
association with long-term changes to our life history cycle and pattern
of growth and development—is important for understanding how our
past hominin ancestors and future extratempestrial descendants could be
part of the same evolutionary history here on Earth.

8.3. Heterochrony and Paedomorphosis in Hominin Evolution

Many different factors have, are, can, and undoubtedly will contribute
to the intricate and uniquely complex process of human biocultural evolution. In addition to environmental change, increased social competition,
and natural and sexual selection, neoteny—meaning the retention of juvenile traits into adulthood—is also important for understanding long-term
evolutionary change in the hominin lineage. Though it should be pointed
out that none of these evolutionary influences are mutually exclusive, as
each have helped shape our past, present, and what is considered to be
our future physical, cognitive, and behavioral state through deep time.

As stated previously, a reduction and retraction of the mid and lower
face occurred in association with encephalization in the hominin lineage.
In fact, our expanding brains would not have had anywhere to go on the

island of the skull if the face had not been retreating throughout this time. This inferoposterior migration of the facial block is related to cerebral development, given the close structural and functional association of these neighboring regions of the skull.[267] More specifically, as the brain and neurocranium continued to expand anteriorly, superiorly, and laterally, the mid and lower face had to accommodate these changes.

This negative correlation between facial projection and neurocranial size/shape is the most conspicuous morphological trend in human craniofacial evolution. It is also easily observable as a static relationship among all mammals. For instance, a larger and more forward-projecting neurocranium is associated with a shorter and flatter face, while a smaller and flatter neurocranium correlates with a longer and more projecting face. This is a ubiquitous bauplan (i.e., an assemblage of morphological characteristics) that holds true across most animal taxa. Additionally, this size/shape relationship is observable as a common structural/functional feature in architecture, aeronautical engineering, and automobile design, considering how difficult and dangerous it would be to mount a truck camper to a corvette, for example.

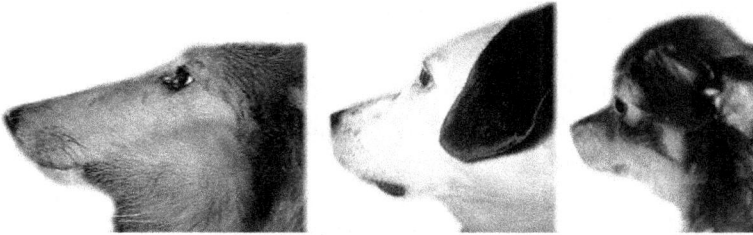

Figure 8-2. *Series of images showing the generalized pattern of craniofacial architecture across three different breeds of domesticated dogs, in which smaller and "cuter" breeds have shorter faces and a more bulbous neurocranium (right), relative to other breeds with flat foreheads and long projecting faces (left).*

This omnipresent association is also readily apparent in domesticated animals, and most notably among domesticated dogs. For instance, picture a Collie sitting next to a Dalmatian, sitting next to a Chihuahua (figure 8-2). This inverse relationship between the size and shape of the neurocranium and face is rather obvious, where a larger and rounder forehead correlates with a shorter face, while a low and sloping forehead is associated with a long and projecting face. This same anatomical tradeoff is even more striking when considering how these traits develop among wild canids, compared to the most highly domesticated dogs of our own creation.

Drawing from these same principles, a graduate school colleague and I developed what we presumed would be a million-dollar idea. It involved manipulating the canine genome to produce an ultra-cute breed that would remain a puppy its entire life. We speculated that we could make a fortune, calling our genetically modified dog breed the *Paedopup*. However, it wasn't long before we realized that this had essentially already been done. These small, flat-faced, big-eyed, big-headed, yappy purse dogs are living proof of long-term genetic manipulation that has resulted in the retention of quintessentially puppy-like traits into the adult form of these animals.

We chose the name "Paedopup" for our visionary late-to-the-game strain of domesticated dogs because 1) they were dogs that would stay pups their whole life and 2) because *paedomorphism* refers to this specific evolutionary and ontological phenomenon in which organisms retain juvenilized characteristics into adulthood. This is a form of neoteny that occurs over evolutionary time and is easily observable across a number of animal species, particularly in highly domesticated ones like dogs and humans.

Neoteny can be somewhat difficult to comprehend, which is likely because it involves changes to ontological time (i.e., rates of growth and development) over evolutionary time, and time itself is a bit mystifying. However, it is actually a relatively straightforward aspect of evolutionary developmental biology that is part of a broader category known as *heterochrony*. Heterochrony involves developmental changes in the rate or timing of events in an organism's life history cycle—meaning its various stages of growth and development—over evolutionary time. Modification to the onset and offset of specific developmental processes results in changes to the size and shape of the organism in adulthood. With regard to paedomorphism specifically, the result of these changes to the timing and rate of ontogenetic events is a fully mature adult that looks much more like the juveniles of its ancestors.

It is helpful to consider this process, and the morphological changes associated with it, in the context of our own species, largely because humans are among the most paedomorphic of all animal species. This evolutionary trend toward increased paedomorphism is manifested in the size, shape, and, perhaps most conspicuously, the relative proportion of different cranial and postcranial features (figure 8-3). In looking at figure 8-3, these changes are readily apparent as you move from the right to the left side of the image, where it can be seen that the individual on the left possesses far more juvenilized traits relative to those toward the right side of the image.

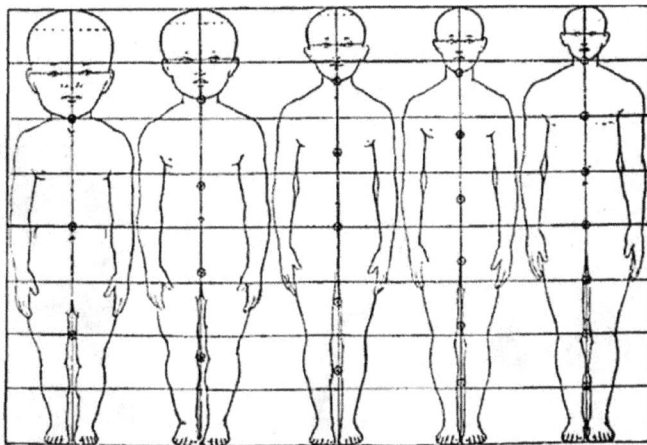

Figure 8-3. *Changes in human cranial and postcranial proportions associated with neoteny.*

Another way of demonstrating paedomorphosis throughout human evolution, which currently only exists as a thought experiment, would be to abduct the most "average" adult human alive today and then take them back 300,000 years into the past. Finding a person that looked the most like them in that period would likely involve searching among the teenagers, as opposed to the full-grown adults of 300,000 years ago. If we took the adult from our time back 900,000 years in the past, we might find they look more like the 10- or 12-year-olds. At 2 million years in the past, we may be better off searching among the 6 to 8-year-olds for someone with the same cranial and facial proportions as our adult modern human.

In essence, the farther back in time we go, the more our adults begin to resemble the younger individuals of those past generations in shape and bodily proportions. A somewhat extreme example of this can also be seen by contrasting the adult human bauplan with that of an infant and adult bonobo chimpanzee (figure 8-4). In this comparison, the adult chimpanzee clearly evokes our ancient ancestral hominid heritage. However, the infant chimpanzee somewhat resembles a tiny hairy human grandmother. So relatively speaking, hominins have retained—while also building upon—this ancestral infantile form, since we last shared a common ancestor with chimpanzees around 6–8 million years ago.

Although neoteny is a very real and easily observable attribute of humans, it is difficult to know at what exact age and during what exact time period one might find a juvenile ancestor whose morphological form is comparable to that of an adult modern human. In order to assess the rate of neoteny throughout the human past, we would first require a large sample size to draw from, with equal representation by both sexes, all age groups, and all geographic variants, drawn from multiple different times across the whole of hominin evolution.

Figure 8-4. *Image illustrating heterochrony, and specifically neoteny, in human evolution, which is apparent from our greater resemblance to a juvenile than an adult bonobo chimpanzee.*

Another problem with hypothesizing about where modern humans fall along the continuum of changing morphogenetic trajectories relative to that of our ancestors and descendants is that heterochronic change is not constant through time. In fact, neotenous traits are more apparent in the skulls of modern humans, even when compared to recent human groups from the Upper Paleolithic only 40,000 years ago. This suggests that the rate of paedomorphosis has been accelerating during recent hominin evolution relative to more distant periods of the past. Therefore, accurately calculating and plotting the speed at which these ontogenetic shifts occur would require many more specimens and much more data than what is currently available to paleoanthropologists.

It is also difficult to determine the exact mechanisms involved in driving this accelerated rate of neoteny in humans. However, a 2005 doctoral thesis by evolutionary psychologist Paul Wehr may provide some insight.[268] Experimental evidence from his study demonstrated that facial underdevelopment was related to the development of progressively larger brains. More specifically, he found that increased orthognathism allows for greater brain expansion and higher cognition in hominin evolution. Though it should be noted that mate choice showed an even stronger association with facial underdevelopment.

Wehr's research showed that craniofacial paedomorphism and a generally more youthful face were significantly positively correlated with perceived attractiveness. Therefore, and as is typically the case with human biocultural evolution, both morphology and mate preference are important for understanding the holistic process of paedomorphosis, as they remain synergistically amalgamated through time. In other words, facial underdevelopment contributes to sexual preference, mate choice, and differential selective fitness within generations, while facial retraction and increased paedomorphosis are also a driving force behind increased cerebral expansion and higher intellect through time. As such, these factors cannot be viewed in isolation when looking at the entire evolutionary picture.[268]

The development of larger more projecting foreheads and smaller cuter faces, which occurred in association with heterochrony and paedomorphosis, has been a prominent and accelerating trend in human evolution. If this pattern of anatomical change were to continue into the future, as it is likely to do, then our adult descendants will retain even more juvenilized traits and to a far greater extent than modern humans, relative to our distant and even recent hominin ancestors. Moreover, because of the neotenous nature of this process, the continuation of these conspicuous craniofacial changes would mean that we may now be hazily gazing upon the morphological form of tomorrow's extratempestrials, in looking at the tiny humans of today (figure 8-5).

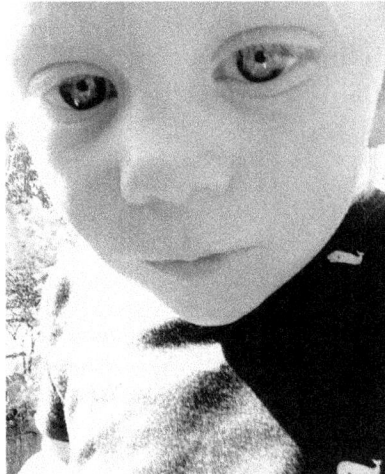

Figure 8-5. *With the continuation of paedomorphic trends into the human future, certain adult characteristics of our extratempestrial descendants (i.e., proportionally large neurocrania, large eyes, and small faces) may be reflected in the skulls of modern infants.*

8.4. Diet, Lithic Tools, Fire, Language, and Love

In addition to bipedalism, social competition, sexual selection, and neoteny, changes in diet and culture were also important to the unique long-term hominin morphological trend of bigger brains and smaller faces. This is largely due to the advent of tools, as well as the introduction of new subsistence strategies and food processing methods allowed for by these tools. For instance, animals that eat hard or coarse foods generally have large teeth, large facial bones, and powerful jaw muscles, which include animals like equids, bovids, gorillas, and an extinct group of hominins known as Paranthropus, or the robust Australopithecines.

Animals that eat softer foods generally don't require large teeth or strong masticatory (chewing) musculature, simply because it takes less effort to get the digestion process started.[269] Furthermore, animals like ourselves that have developed the ability to cut and cook their food require even less in the way of teeth and facial musculature because, as stated by anthropologist Shara Bailey, these tools were like "external teeth." In an interview for National Public Radio, Bailey goes on to say that "Your teeth are really for processing food, of course, but if you do all the food processing out here (with tools), if you are grinding things, then there is less pressure for your teeth to pick up the slack."[270]

In a recent study published in the journal *Nature*, Katherine Zink and Daniel Lieberman from the Department of Human Evolutionary Biology at Harvard University, were able to demonstrate just how much we have outsourced chewing to external tools in human evolution. Their biomechanical research used experimental methods to show that if meat made up about one-third of *Homo erectus'* diet, slicing and pounding the meat with early stone tools could reduce the average number of chewing cycles by 13%, or the equivalent of about 2 million chewing cycles per year. By placing surface electromyography (EMG) electrodes on the skin overlying the muscles most involved in mastication, they found that the overall force required to chew uncooked meat would have been reduced by about 15% after it was processed using Lower Paleolithic tools.[271] With a reduction in the bite force and number of chewing cycles required to process meat—and particularly in places where fire was not yet common—these lithic tools would have helped reduce the need for larger teeth, meaning more space became available for our brains to continue growing larger in the skull.

In addition to tooth size, a relationship also exists between tooth number and facial size/shape in mammals. In fact, it is becoming increasingly common among modern human groups for our third molars (M3), or wisdom teeth, not to develop in the mouth at all. Our early hominin ancestors consistently had three molars in each quadrant of the mouth, for a total of 12. However, a random mutation that suppresses the formation of M3 arose more recently, with the earliest fossil evidence of this observable in a human specimen from China that dates to 300,000–400,000 years ago.[272] Since this time, the mutation has spread and increased in frequency.

The rate of third molar agenesis also tends to be higher among modern human populations with generally flatter faces. For instance, African-Americans have a craniofacial structure that is somewhat more prognathic compared to other ancestral groups, and about 11% of individuals of African descent lack at least one of their third molars. European-Americans exhibit a moderate degree of facial orthognathism and are missing at least one of their wisdom teeth at a rate of about 10–25%. While individuals of East Asian descent, from the region where this mutation is thought to have originated and who are generally more orthognathic by comparison, lack at least one M3 at a frequency of about 40%.[272]

This clear and persistent trend toward reduced tooth size and number, in association with a reduction in masticatory musculature, indicates that we began to eat meat to a far greater extent beginning around 2.5–3 million years ago.[273] This is also apparent from looking at changes to our postcranial anatomy, including a prominent trend toward a reduction in our thoracic anatomy, which began with our gracile Australopithecine ancestors around this same time. But perhaps the most notable change that occurred in association with our changing diet, and increased reliance on meat specifically, was the origins of the most conspicuous trend in hominin evolution, the development of larger and more complex brains.

Compared to course vegetative materials like fibrous leaves, stalks and roots, it takes a lot less gut to digest meat. This is a part of why carnivores tend to be slenderer, while leaf-eating species have larger potbellies and oftentimes multiple stomachs. Because our diet shifted more toward meat, we started to develop a taller and narrower body build. Additionally, it is generally accepted among anthropologists that this reduction in our gastrointestinal system, which occurred in association with more animal products in our diet, helped meet the increased metabolic needs of our expanding brains. In this way, we were able to grow larger brains at the expense of an energy-intensive gastrointestinal system.

This evolutionary tradeoff is part of a prominent anthropological theory known as the *expensive-tissue hypothesis*.[274] According to Leslie Aiello, director of the Wenner-Gren Foundation, and one of the anthropologists who helped develop this idea:

> You can't have a large brain and big guts at the same time. Digestion was the energy-hog of our primate ancestor's body. The brain was the poor stepsister who got the leftovers. What we think is that this dietary change around 2.3 million years ago was one of the major significant factors in the evolution of our own species.[270]

After we began to focus more of our energy on obtaining meat, a number of important changes to our guts, masticatory anatomy, brains, and even social systems started to take place. In the early stages of this shift in subsistence strategy, we were undoubtedly forced to rely on scavenging to procure meat, as we had not yet developed the tools, intellect, or communication skills to facilitate a successful hunt. It is hard to know when we stopped scavenging and started to focus on hunting for our own meat, but a conservative estimate would put it with *Homo erectus*, likely sometime around 1.5 million years ago. This is indicated by the advent of more sophisticated stone tools, known as the Acheulean tradition, as well as an increase in the size of Brodmann's areas 44 and 45, which is the location of Broca's area, an important speech center in the brain.

This linguistic brain region began to appear more prominently in the endocasts of *Homo habilis* around 2.5 million years ago[275] and to an even greater extent in *Homo erectus* beginning around 1.8–1.5 million years ago.[276] Expansion of these linguistic areas of the brain may have been associated with the need for better communication prior to and during the hunt. Additionally, an intricate link may exist between the cognitive processes associated with language and the production of stone tools,[277] particularly as it relates to handedness and lateralization of the hominin brain.[278]

The manufacture of lithic tools, the onset of widespread hunting, and the equally momentous advent of fire, were tremendously important changes that helped accelerate the rate of facial orthognathism, cerebral expansion, and cognitive development during the time of *Homo erectus*. Together, these sit atop an exclusive list of momentous hominin achievements that helped advance and accelerate human biocultural evolution through time, which also includes the advent of agriculture, harnessing fossil fuels for energy, and the invention of electronic devices such as computers and mobile phones.

These novel additions to past hominin lifeways were important, both individually and in conjunction with one another. For instance, hunting provided more opportunities to procure meat as compared to scavenging; it facilitated better communication; and being the first one to a kill meant less competition from large and dangerous predators, as well as hungry and potentially hostile neighbors. Additionally, more meat meant softer food, which allowed our faces and guts to get out of the way of our expanding brains. The advent of fire and cooking made the meat safer, easier to digest, and softer still,[279][280] while the development of new and better tools to cut and tenderize meat further reduced our masticatory anatomy and allowed us to continue growing even larger brains. [271][281][282][283] This trend in hominin evolution continued with the advent and spread of agriculture, which brought forth new foods and technologies that eased the mechanical demands of chewing, furthering the progression of these same morphological changes to the modern human skull.[284]

8.4.1. Fire and Society

The earliest evidence of hearths, which indicate controlled use of fire, cooking, and a centralized gathering place for shared warmth and light, are associated with the Lower Pleistocene sites of Koobi Fora, located near the eastern shore of Lake Turkana in Kenya. Materials associated with these early hearths have been radiometrically dated to 1.6 million years ago.[282][285][286] Interestingly, this coincides with the increased use of flint in tool manufacturing, as well as with generally drier conditions and more bush fires in this area of eastern Africa where fire use is thought to have originated.[285] This suggests that sparks produced while flint knapping could have occasionally created unintentional fires, and with them, the realization that fire could be made with relatively little effort.

As mentioned previously, this is also considered to be an important time for the development of more complex language in hominin history. The pressures of circumstances unique to our ancestors are thought to have helped shape an early protolanguage beginning as early as 2–4 million years ago.[287][288] This protolanguage then began to advance with *Homo erectus* around 1.6 million years ago and eventually reached its current level of complexity with *Homo sapiens sapiens* in the recent human past.[289]

It is easy to imagine how tactical group-hunting could foster more complex linguistic skills, although fire likely played a prominent role as well. For instance, the centralization of people around the warmth and light of a fire may have acted as a linguistic petri dish, providing a consolidated competitive social environment where the best orators could vie for the approbation of their peers through verbal discourse (figure 8-6). This competitive linguistic arena would be expected to further catalyze cognitive and communicative competency through time.

Winning the favor of others, and especially those we are attracted to, takes a bit of cunning and a fine-tuned set of culturally dependent skills. These are critical components of the enculturation process and require a lengthy period of trial and error, which often involves some embarrassing social situations. However, these are vital to shaping successful future encounters through the process of learning. Because of the centralizing role of fire, this could have been an important place for the younger members of society to hone skills that would someday become crucial in their efforts to garner the respect of their peers, broker economic agreements, forge political alliances, and perhaps, most importantly, vie for mates.

Figure 8-6. *Artist depiction of an Upper Paleolithic scene where lively oral discourse transpires near the light and warmth of a nocturnal fire.*

Social fires are an ideal venue for boasting of hunting success, trading war stories, probing knowledge, strategizing, passing down oral histories, and where eloquent oration may help some individuals achieve a higher selective fitness. In these ways, fireside forums were likely an important component of our individual and species-wide intellectual journey, as this intra-generational competition and sexual selection for higher intellect would also be expected to contribute to neurocognitive development and cerebral expansion in a long-term evolutionary capacity.

8.5. Traces of the Past Trend toward the Future

Numerous anatomical changes characterize the long history of hominin evolution, although habitual bipedalism, reduced facial prognathism, and cerebral expansion most define our lineage. These key morphological trends have occurred in concert with one another and in association with vast changes to our geographic, ecological, sociocultural, linguistic, and physical environments. In spite of the tremendous variation experienced by our hominin ancestors, these same morphological trends have persisted, as our faces have continued to shrink in association with our expanding brains.

It is impossible to know what specific factors might shape the future of hominin evolution or how these could affect our biology and culture going forward. However, given the veracity with which these unwavering 6-million-year-old trends have persisted, certain predictions regarding our descendant's general craniofacial bauplan may be less speculative. Considering only these most enduring trends throughout the hominin past, our bipedal hominin descendants are likely to possess an even rounder neurocranium, situated above even smaller faces, which exhibit an even greater degree of paedomorphosis. With this in mind, one can appreciate how the vast majority of reports of close encounters, which describe alien beings with these same craniofacial characteristics, are quite simply us, as more evolved members of the same hominin lineage.

9

Astrophysical Anthropology: Linking
Human Biocultural Evolution through
Space and Time

*If I had been asked, I would have said that the nature of my experience indicated
that the visitors hadn't been here too long, and that I had been studied by a team of
biologists and anthropologists.*[290]

– Whitley Strieber

9.1. Probing the Future via Our Communal Pasts

A pattern represents any discernible regularity of things that repeat
in a predictable manner. When the same patterns occur with prodigious
regularity over long periods of the past, more precise predictions about
future outcomes can be made. Generally speaking, the same craniofacial
trends have endured and accelerated throughout hominin evolutionary
history, in spite of diverse and often divergent external circumstances. In
fact, the persistence of these trends across highly capricious environments
points to the adaptive value of these unique features, as well as the likelihood that they will persist long into the human future.

For instance, bipedalism, free use of our hands, increased meat consumption, fire, tools, big intelligent brains, and the shrunken faces that go
with them, have allowed for greater adaptability in shifting environments
over the last two million years of human evolution. Furthermore, our
environment continues to change and is likely to do so with even greater
ferocity due to recent anthropogenic factors (climate change, sea level rise,
increased forest fires, etc.). As a result, these same accelerating patterns of

anatomical and cultural change—and the benefits of behavioral flexibility that they impart—should extend into the future of hominin evolution, albeit with some level of geographic and temporal fluctuation throughout.

As an example, even though our cerebral and neurocranial anatomy have grown consistently larger throughout the vast majority of human evolution, the size of our brains has actually been decreasing over the last 10,000 years. This change is largely the result of a broader reduction in cranial, facial, and skeletal size, in which the brain, cranial vault, oro-facial skeleton, and overall skeletal robusticity have decreased across most geographic groups since the Holocene and even earlier in some regions of the world.[291][292][293][294] For example, a 10–30% reduction in craniofacial dimensions occurred among modern humans in Western Europe since the Upper Paleolithic, which began about 35,000 years ago in this region.[295]

Looking across the entire history of hominin evolution, it is likely that this recent decrease in brain size is more of a diachronic secular trend, rather than a reversal of this persistent pattern of encephalization over the last 6 million years of human evolution. In fact, when averaged across even the last 500,000 years, it is clear that our brains have experienced an extraordinary increase in size, which has also occurred in association with a marked change in neurocranial shape. Most notably, we have evolved proportionally larger and more anteriorly positioned temporal poles, an expanded precuneus and parietal surface, and a much wider and more anteriorly positioned frontal cortex, which have contributed to the uniquely globular, or bulbous, craniofacial shape that defines us as anatomically modern *Homo sapiens*.[296][297][298][299]

Even though brain size has decreased during the Holocene, some research suggests that the accelerated rate of facial reduction and increased neurocranial globularity are likely to persist and may even accelerate further. As will be discussed in subsequent chapters, this prediction stems from a number of past and ongoing biocultural trends, such as a more rapid rate of cultural and technological change, a continued cross-cultural preference for youth and paedomorphic beauty, a recent pattern of self-domestication, a dominant trend toward craniofacial feminization, and perhaps most notably, a recent increase in the rate of cesarean sections, which may aid in circumventing one of the last major impediments to further cerebral expansion in hominin evolution.

If our culture and technology continue to progress while we keep developing more balloon-headed and child-faced characteristics, then we may conceivably be on a path toward becoming the same outwardly advanced extratempestrials reportedly encountered by countless people across the world and through time. Without publicly available pictures, videos, a fully articulated skeleton, or, ideally, a living, breathing, wholly animated future human, it is undoubtedly more challenging to put together the disjointed pieces of this temporal puzzle.

Although, as demonstrated earlier in this text, countless individuals of sound mind assert that they have experienced tangible forms of contact and communication. They have provided exhaustive accounts of the behavior and physical form of these extratempestrials and their IFOs, as well as detailed descriptions of events that transpired as part of these extraordinary encounters. These accounts certainly cannot provide the same level of understanding as that which could be attained by means of a real-time biomedical examination of an actual living specimen, which would seem to be their preferred method of investigating hominins of disparate times. However, a detailed analysis of exemplary abduction reports and other instances of intimate contact, may offer some insight into the future form of our species.

9.2. Abduction Reports as a Latent Look at the Future of Hominin Evolution

Although alien beings are portrayed on television and in movies in a number of strange forms, there is actually a very high degree of similarity in the descriptions provided by individuals who claim to have had direct contact with them. Reports that rank highly on Dr. J. Allen Hynek's *Strangeness and Probability Scale* could be instrumental in assessing whether the observed physical and cultural traits of these "aliens" are consistent with what we would expect to see in our distant human descendants, if past biocultural trends were to continue long into the future.

Given the antiquity and global prominence of this phenomenon, most people have at some point in their lives seen images depicting the quintessential form of the aliens described by those who have had the opportunity to observe them up close. However, for those who may still lack a visual comprehension of this archetypal alien form, an artist's depiction is provided in figure 9-1, and more detailed descriptions of vetted abduction accounts can be found below. Despite not having any ears, among other potential inconsistencies, the large bulbous head, small childlike nose, mouth and chin, large eyes, and generally gracile feminized form are extremely common characteristics in reports of abductions and other close encounters.

Figure 9-1. *Artist's depiction of the archetypal physical form of extratempestrials*

In the book *Communion: A true story,* author Whitley Strieber describes a personal encounter that occurred on the evening of December 26th, 1985, in which he was abducted from his cabin in upstate New York by what he referred to as "visitors."[300] Strieber provided details regarding a few different interactions that took place between himself and the extratempestrials, both in his cabin and aboard their IFO. While on the ship, Strieber—like so many others—found that he was the subject of a physical examination.

> It appeared to be a small operating theater. I was in the center of it on a table, and three tiers of benches were populated with a few huddled figures, some with round, as opposed to slanted eyes . . . These had wide faces, appearing either dark grey or dark blue in that light, with glittering deep-set eyes, pug noses, and broad somewhat human mouths.[301]

Strieber goes on to describe additional individuals within the room who seemed to have the same but even more pronounced physical features as those seated around the operating amphitheater:

> The most provocative of these was about five feet tall, very slender and delicate, with extremely prominent and mesmerizing black slanted eyes. This being had an almost vestigial mouth and nose. The huddled figures in the theater were somewhat smaller, with similarly shaped heads, but round black eyes like buttons.[301]

During a hypnosis session conducted on March 1ˢᵗ 1986 by psychiatrist Dr. Donald Klein, Strieber again recalled these same encephalized and overly ocular characteristics, stating that "...it had a bald, rather largish head for someone that size. And that its eyes are slanted, more than an Oriental's eyes."[302] This account of the physical form of these individuals is remarkably consistent across abductee reports. Furthermore, the physical traits described are also consistent with what we would expect to see in the morphological form of our distant descendants, based on prominent long-term trends in our cerebral, ocular, cranial, facial, and postcranial anatomy throughout the last 6 million years of hominin evolution.

This is corroborated by a 1957 abduction report provided by a respected Brazilian lawyer named Antonio Villas-Boas. Although the extratempestrials he encountered were entirely similar with regard to their physical form, he reports having much more intimate contact with one of his captors, who he described as female with blond hair, blue slanted eyes, a wide face, thin lips, and a very pointy chin.[303]

Strieber also makes reference to the Villas-Boas abduction account in his book, and even acknowledges her human-like qualities, stating that the female observed by Villas-Boas "sounds very much like a cross between the individual I saw so clearly as the eidetic image, and a human being."[304] Multiple other eyewitness accounts also emphasize the similarity between humans and extratempestrials, which lends support to this proposed phylogenetic link between us now and us in the distant future.

Such accounts may also provide some insight into the potential purpose of these visits to the past. For instance, as per Whitley Strieber's quote that begins this chapter, these narratives often point to an anthropological and biomedical research objective. Among other markers, this is indicated by their sampling procedures, major research instrumentation, propensity to sedate or tranquilize their research subjects prior to examination, their placement of abductees on examination tables, a formal doctor/patient type of interaction, their language and description of procedures, as well as the way these abductions unfold, which resembles what modern anthropologists would do while examining past hominins, if we too possessed the capacity for backward time travel research.

To further illustrate this common biomedical research component of extratempestrial abductions, as well as the ubiquity of reports involving descriptions of human-like aliens, an exhaustive query of vetted accounts archived by the *UFO Casebook's* Alien Abduction Case Files was carried out.[305] Of the many cataloged encounters that occurred between 1950 and 2005, five were selected that point to the continuation of past morphological and cultural trends in hominin evolution, and another five accounts were selected among those demonstrating a focus on human and non-human animal research.

9.2.1. Abduction Reports Indicating the Continuation of Morphological Trends in Hominin Evolution

South Ashburnham, Massachusetts, U.S., 1967, Betty Andreasson Abduction[306]

...about five-feet tall. The other four appeared to be about a foot shorter. All of the beings had a pear-shaped head, with wide eyes, and small ears and noses. Their mouths were only slits, and never moved.

Buff Ledge Camp, Vermont, U.S., 1968, Michael Lapp and Janet Cornell Abduction[307]

...transparent dome occupied by two childlike creatures...having elongated necks, big heads, and no hair. Their eyes were also large and extended around the side of their heads ... large eyes, a mouth without lips, no ears, and two small openings for a nose.

Apache-Sitgreaves National Forest, Arizona, U.S., 1975, Travis Walton Abduction[308]

... beings were a little under 5 ft. in height, with a basic humanoid appearance. ... Totally bald, their heads were disproportionately large for their little bodies ... incredible eyes! Those glistening orbs had brown irises twice the size of those of a normal human's eye, nearly an inch in diameter! The iris was so large that even parts of the pupils were hidden by the lids, giving the eyes a certain catlike appearance. There was very little of the white part of the eye showing. They had no lashes and no eyebrows. Their little mouths never moved ...

Emilcin, Poland, 1978, Jan Wolski Abduction[309]

They were similar to us thought [sic] tiny, delicate, of small height —about 150 cm. [5 ft.]. They had slightly slanted eyes. ... They cheekbones were protruding. Once I saw Chinese men and they [the beings] slightly reminded them. But Chinese seemed to more massive and taller There were no brows visible ... I don't know what they had in fact—some make up or masks. They only got slanted eyes. I could see some white in their eye corners. And their teeth were also white in color.

Sydney, Australia, 1988, Peter Khoury Abduction[310]

These two women looked human in nearly every way. They had well-proportioned adult bodies. One looked somewhat Asian, with straight dark shoulder-length hair and dark eyes. The other looked perhaps Scandinavian, with light-colored ('maybe bluish') eyes and long blond hair that fell half-way down her back ... these women were not exactly human. Their faces were somewhat odd—not unattractive, but too chiseled, with very high cheekbones and eyes that were two or three times larger than normal.

9.2.2. Abduction Reports Indicating an Emphasis on Future Research into the Hominin Past

Aldershot, Hampshire, United Kingdom, 1983, Albert Burtoo Abduction[311]

They were about four-foot-high, dressed in pale green coveralls from head to foot ... One of the beings told the old man to stand beneath an orange light, which appeared to scan him for a few minutes. 'What is your age?' asked the entity, in a 'sing-song' voice which sounded like a mixture of Chinese and Russian. When he replied that he was 78, it declared: 'You can go. You are too old and infirm for our purposes.'

North Canol Road, Northern Canada, 1987, Kevin Abduction[312]

I had been looking into the eyes of a grey type creature. I could hear in my mind a voice saying, 'there is nothing to worry about.' I could hear him talking in my mind ... There were 3 or 4 of these types walking around, but only one talked to me ... I then sat up and had an idea of what might be going on and I asked, 'Are you going to do experiments on me?' and the one said, 'They've already been done.' I felt really good then because, except for a strange sensation in my hands everything felt normal.

Near Edinburgh, Scotland, 1992, Garry Wood and Colin Wright Abduction[313]

He [Colin] was stripped naked and placed unresisting in the chair and subjected to some form of non-intrusive physical examination. This memory segued seamlessly into being naked in a transparent container made from a material rather like glass or Perspex ...

Outside the container he could clearly see other men and women, all naked and all in transparent containers like his. He also saw a number of tall, humanoid creatures ... an angular device rose from the floor ...The entire machine moved up and down continuously and the appendage swung from left to right; although there was no pain, Colin thought it might be scanning him ...

Like Colin he [Garry] described being in a featureless circular room lying on a flat table, he was unable to move although he does not recall being physically strapped down ... A tall, incredibly thin frail looking creature slowly, almost painfully, emerged. Although bearing a marked similarity to a traditional 'grey' it appeared emaciated, like a skeleton covered in skin. Bizarrely, he also recalled a small man apparently quite human dressed in a neat black suit complete with collar and tie who was watching the proceedings. He was standing among the entities all of which seemed quite deferential towards him. In all, Wood remembers there being around 20 or 30 creatures present, the majority tall, a pallid grey colour and frail looking.

Gundiah-Mackay, Australia, 2001, Amy Rylance Abduction
(witnessed by 2 others)[314][315]

...remembered waking up lying on a bench in a strange rectangular room ... Soon an opening appeared in the wall and "a guy" about 6 feet tall walked into the room.The man appeared to be slender in build but in perfect proportion, covered head to foot in a full body suit. He had what seemed to be a black covering mask on his face, with a hole for his eyes, nose and mouth. Amy felt she had been there a while. The guy told her they were returning her to a place not far from where they took her from, because the lights were wrong at the property and it wasn't safe.[316]

Approximately ninety minutes after Petra witnessed the supposed abduction, Amy was found disorientated and covered in mud at Mackay Queensland ... about 790 kms from Gundiah.This trip, when driven by car, would usually take at least 8–9 hours. On discovery Ms. Rylance stated that she had been abducted for several days. When examined by Mackay hospital staff they found that she had in fact not eaten for days, and her usually shaven bodily hair had grown considerably.[315]

Florida, U.S., 2003, Anonymous[317]

> The room was pure white with what I think was fluorescent lights all over the place. I was laying on a hard surface . . . I opened my eyes and there was what looked like a human female only taller than anyone I've seen standing above me. Their skin was rough and cold and their eyes looked like they saw everything and knew everything . . . I saw more movement in the background and voices talking in a language I couldn't understand . . . I then remember waking up on a hard-white bed. I saw a few cages in the room with small animals in it (squirrels, rabbits etc.) Once again I heard murmurs in another language, then it sounded like they said it inside of my head but a soft voice whispered, 'We're taking you home.'

9.3. Constructing the Other

These and other abduction accounts provide a glimpse into the oddities encountered while interacting with individuals from another time. They also make it easier to understand how contemporary humans, and our technology, may be misunderstood by our own distant ancestors, if we too were able to return to a former time to study them. For instance, if modern humans boarded a spinning disc and traveled back 1.25 million years into the past to abduct our *Homo erectus* ancestors, they would undoubtedly describe us as having a large bulbous cranium, slender bodies, a small nose and mouth, and a pointy chin. Additionally, depending on the sex, age, and ancestry of the individuals from our own time, these *Homo erectus* abductees may also describe us as having large and slanted eyes, very little body hair, and potentially completely bald heads.

We would also likely be observed descending from the sky, which could lead our distant ancestors to falsely assume that we came from a different planet rather than a different time. Though considering the morphology, language, and technology of extratempestrials, the vast majority of evidence suggests that they, like us in this notional scenario, are actually from time, as opposed to space. In fact, with regard to the above abduction reports, the astute reader may have noticed that Amy Rylance was discovered about an hour and a half after her housemate witnessed her being abducted through a window. However, Amy reported that she had actually been gone for multiple days, which the Mackay hospital staff corroborated. Such an anomaly could only occur in the presence of a time machine.

Contemplating how our ancestors may view us if we were the ones conducting research in their time helps conceptualize the oddities, yet also the similarities, in the morphology and culture of our extratempestrial descendants. Though an important difference between how they would view us and how we should view our distant descendants is that modern humans are now aware of our evolutionary ancestry, and as such, we have more knowledge and tools available to help piece together these shared phylogenetic relationships through deep time. However, in spite of our current knowledge of hominin evolution and universal descriptions of extratempestrials as "humanoid," "human-like," and even "human in nearly every way",[310] we continue to think of them as extraterrestrial beings from somewhere else in the universe.

It is somewhat astonishing that people who have observed extratempestrials don't instantly recognize them as members of our species, or at least members of the hominid clade, given that they also lack a tail, among other shared derived characteristics. In fact, it would seem as though we share far more derived traits with extratempestrials than any other ape, including bipedalism, the one trait that defines our hominin lineage above all others. Even for those of us who have not been visited by our distant progeny or who have not had the opportunity to view their physical form directly, descriptions provided by those who have overwhelmingly point to something that is undoubtedly primate, and unmistakably hominin.

As early as the 17th and 18th centuries, it was apparent to naturalists such as John Ray, Georges Cuvier, Comte de Buffon, and Carolus Linnaeus that humans resembled apes more than other mammals, based on a host of shared traits. This association was apparent, despite the fact that these early naturalists still lacked any understanding of true taxonomy that is based on evolutionary relationships and common ancestry. Though even with our modern knowledge, we continue to overlook what seems to be an obvious taxonomic relationship between ourselves and these slightly more derived humans with whom we share even more common characteristics, resulting from an even deeper shared ancestry.

We now possess a much more profound knowledge of evolutionary change and the forces that helped shape modern human variation. With this current awareness, one would think we would be quicker to recognize the true origins of those visiting us from an even more advanced human future. Though admittedly, history has shown that even recent human groups struggled with recognizing other humans as human, particularly during the early stages of contact and colonialism, when people began encountering disparate groups from other geographic areas.

Recognizing the extratempestrial phenomenon as an informative link between different times—as we eventually began to do with people from different places—may shed some light on the otherwise opaque future state of our species. More specifically, observing them as they observe us could help answer some rather complex questions, such as whether modern human variation and geographical races will persist into the future, how these relate to temporal variation resulting from further evolutionary change, what are the geographic and/or temporal origins of these extratempestrial researchers, how does language change through time, what principal forces may help shape our future biology and culture, and as discussed previously, what characteristics of their technology makes this intertemporal interaction possible.

9.4. Human Variation—Past, Present, and Future

The physical traits of extratempestrials observed during close encounters are remarkably similar to those of hominins, which implies genetic continuity between our past and future. Abduction reports and other instances of close encounters also indicate that some level of physical variation exists among extratempestrials. This is understandable given the nature of time travel and its potential to add temporal variability beyond the sex, age, and geographic variation that is already present in and among human populations from contemporaneous periods. In other words, in the presence of a time machine, we would expect to see additional human variation among living individuals representing different stages of hominin evolution. Additionally, if there is to be collaboration among researchers from different periods, this would be expected to add to the range of physical variation observable among extratempestrials together at any specific time.

This is to say that individuals representing different stages of our future evolution could be present aboard the same craft at the same time. In this way, intertemporal collaboration among individuals from different periods would further contribute to the observable variation among extratempestrials seen in our time, at any given time. Many modern research endeavors—such as the Large Hadron Collider (LHC) experiments at CERN for example—involve collaboration among individuals representing multiple different countries, who exhibit a relatively high degree of ancestral geographic variation. In the future, intensive investigations of the past may also involve collaboration among individuals representing different geographic regions, as well as different time periods, meaning research teams aboard an IFO could potentially comprise individuals from many different eras, perhaps separated by as much as hundreds, thousands, or even tens of thousands of years.

Modern humans exhibit some degree of variation in height, weight, skin color, head shape, eye shape, etc. This broad range of physical variation exists in spite of the fact that we have recently become one large interbreeding population. Though even before this, our species already exhibited a somewhat uniform genome[318] and a similar overall craniofacial morphology.[319] The more overt physical differences among human groups, which make us seem more different than we actually are, largely resulted from localized populations adapting to different environmental conditions over the last 200,000 to 500,000 years. Though even considering this diversity among formerly dispersed geographic groups, compared to most other animals, the amount of variation among living humans is actually quite minimal.[323]

Thus far in human history, we have not yet had to consider temporal disparities as a factor contributing to human variation among living populations. This is simply because we have not yet developed the technology to facilitate time travel, which would bring together past and modern individuals with different scales of the same evolutionary characteristics. It is also because we have thus far regarded the IFO phenomenon in the context of extraterrestrial life, without considering that extratempestrials would be able to link our future to various periods of their respective past. Though if people at any point in time become able to visit different periods of the past or future, we must all reevaluate human variation in a much broader context, even if we weren't the ones who developed this technology. With this in mind, we are likely seeing different individuals from different points in time, each time extratempestrials return to study us in their past.

Once a time machine is invented, this technology would be expected to persist throughout the ensuing future. Therefore, it is not unreasonable to suspect that the individuals we observe "now" may have returned from different periods spread widely across that future. Theoretically, an individual from our own time could be picked up for study one week by a group of anthropologists/biologists living 20,000 years in the future and then later that same week by researchers originating from 140,000 years in the future. In the same way that we never stopped using tools and fire once we realized their importance to our existence, it is also not likely that we would simply abandon a technology that allows us to visit the past.

This long-term evolutionary aspect of time travel is also important to consider with regard to the problem of designating a type specimen among our descendants. In other words, it would be hard to determine an exact physical form representative of all extratempestrials, if we are indeed being visited by individuals from multiple different time periods. This aspect of temporal divergence may also help explain reported variation in the height, weight, skin color, and craniofacial form of extratempestrials, particularly those describing more insect or reptile-like individuals, who may have originated from a very distant point in our evolutionary future.

9.5. Extratempestrial Phenotypic Variation—Geography, Ancestry, and Time

Many accounts provided by people who claim to have been abducted by extratempestrials describe their abductors as looking different from one another. This inter-alien variation is often equated to modern geographic racial groups, which could be a fair interpretation, considering that geographic variation exists among extant humans and could persist into the future. Although, over the last 500 years or so, the human population has become much more integrated, and there are indications that we could become even more racially homogenized in the future. Because of this recent trend, we must consider whether these physical differences among extratempestrials reflect disparate intertemporal ancestries as described above, if they are the result of non-heritable acclimations to specific future environments, if they are related to intentional genetic engineering to avoid human population homogenization, or whether they are the result of contemporary geographic variants persisting into the future.

Based on recent and ongoing trends of intercontinental travel and interbreeding among previously isolated populations, it is possible that human geographical differences will continue to wane. Additional blending of traits that developed in specific geographic areas, as a result of adaptations to local environmental conditions, could lead to further homogenization of the human genome. In fact, with rampant gene flow and cultural transmission across the entire human population, it may be inevitable that we amalgamate over time and eventually become a monochromatic and monolinguistic species.[320]

9.5.1. Skin Color

Currently, human skin color varies rather prominently from light to dark, which is the result of past human groups adapting to different ultra-violet light levels at different latitudes and altitudes, over many thousands of years. Abductees often describe their future human hosts as having light, often greyish-looking skin, which could be an indication of contin-ued human genetic homogenization into the distant future. For instance, if everyone on Earth were to suddenly begin randomly mating, it would not take long before the average human skin tone turned to something of a light brownish/greyish hue. This of course assumes that past selec-tive forces related to ultraviolet radiation, vitamin D synthesis, and folate deficiencies, which led to skin color differentiation in humans, would still be mitigated by cultural means.

The existing trend toward homogenization of melanin-levels, resulting from gene flow across different geographic groups, would be expected to continue as a result of even greater cultural integration and increased speed and ease of international travel. Skin color homogenization seems inevitable over the long term and may represent yet another biocultural trend linking our evolutionary past to that of our future. Additionally, in the context of biocultural evolution, light-skinned people valuing dark-er skin, and dark-skinned people valuing lighter skin, indicates that the cross-cultural ideal of beauty is somewhere in between.

However, intentional whitewashing across various forms of media, a multibillion-dollar skin-whitening cosmetics industry for dark-skinned people, a 3-billion-dollar tanning salon industry for light-skinned people,[321] and the false ideals of beauty that go with these, are largely perceived as being socially harmful. Furthermore, many cultural norms associated with the desirability of certain skin tones originated with, and were perpetuated by, European colonialism. In fact, when viewed holistically, these industries and practices represent a form of continued cultural imperialism that re-capitulates 500-year-old racial hierarchies established during the colonial period, which were sustained through forced assimilation, institutionalized racism, and class stratification. According to Tansy Hoskins, author of *Stitched up: The Anti-Capitalist Book of Fashion*:

> The cosmetics industry has traditionally relied on convincing peo-ple that they are incomplete without a particular product. Yet, unlike makeup or fake tan, skin-whitening creams base beauty on a racial hierarchy, fueling intolerance and causing serious social harm.[322]

This is also evidenced by a recent tendency to "whitewash" darker-skinned models in order to make them look whiter. In fact, a rather overt example of this practice caused controversy in 2016 when a *W Magazine* cover was edited prior to publication in order to make two actresses —Zendaya Coleman and Willow Smith—appear much whiter than they actually are.[323] This legacy of European colonialism is also apparent in the film, movie, and music industries, since the respective ancestral group of current artists often had lighter than average skin tone. According to sociology professor Tanya Golash-Boza:

> Colorism is evident in the U.S. entertainment industry, especially for women of color. Many of the most prominent Latina stars—Selma Hayek, Jennifer Lopez, and Eva Longoria—are very light-skinned. The same can be said for prominent female African American stars such as Halle Berry and Beyoncé, who are both very light-skinned.

Certainly, tanning and skin-lightening creams are not going to modify our genome and result in a medium-shade heritable hue for the human population. Additionally, intra-cultural and inter-cultural beauty trends come and go and do not tend to impact long-term human evolution in the same way as environmental forces. However, the current desirability of a skin tone that is somewhere between light and dark, coupled with a global population that is becoming more integrated and interbred, together with the continuation of a roughly 20,000-year-trend toward self-domestication, may move the global human population toward a more homogenized pigment-deficient greyish-brown skin color.

For instance, in addition to these cultural and population-integration factors, which could result in an averaging of global human skin colors, the process of domestication—which is a unique form of self-domestication in humans—also leads to depigmentation over time. Originally demonstrated in foxes, the underlying mechanism of this depigmentation is a retardation in the proliferation and migration from the neural crest of melanoblasts, which are embryonic precursors of the melanocytes that form melanin in the skin.[324 325 326 327 328]

While skin color homogenization across modern human populations would be expected to result in a medium brown hue, this tendency toward depigmentation in highly domesticated species like *Homo sapiens*, could help explain the greyer skin color that is so commonly reported in instances of close encounters. Taken together, recent trends toward worldwide genetic homogenization among previously isolated populations, a cultural emphasis on darker skin for light-skinned people, and lighter skin for dark-skinned people, as well as a dominant trend toward increased self-domestication in *Homo sapiens*, may shift the human population toward a light-brown/grey skin color in the future.

9.5.2. Stature

In addition to skin tone, stature is another physical characteristic of extratempestrials that is somewhat variable across abduction reports. Some have recalled them being around 4 feet tall, while others state that they were upwards of 7 feet in height. In spite of this reported variation, these body height estimates are still consistent with the overall stature of modern and past humans, particularly considering that stature is among the most variable traits among contemporaneous, historical, and evolutionary human groups. In fact, the average height of a population can vary markedly over even short periods of time. For instance, among the Dutch, the tallest people in the world, between 1858 and 2009 the height of boys increased from 163 cm to 183.8 cm, which is an average increase of 21 cm, or 8.27 inches.[329]

This high level of stature variability is largely because height does not carry a high heritability in humans, meaning that it is not completely dictated by our genes. Rather, the average height of individuals at any given time, both within and among populations, is primarily the result of secular trends influenced by diet, health, climate, occupation, income, education, and numerous other cultural and environmental factors.[330 331 332] Regarding the aforementioned Dutch study, an increase in the education level of parents and children was found to be among the most influential factors in growing the average height of individuals in this large study sample.[329]

Long-term variation in human height is also apparent from looking at the fossil record of different hominin groups. For instance, in 2003 a closely related human sub-species known as *Homo sapiens floresiensis* was discovered on the island of Flores in Indonesia. What was most intriguing about these individuals, other than the fact that they existed as recently as 50,000 to 60,000 years ago,[333] was that they were only about one meter tall. In fact, using femoral length estimation formulae for height borrowed from human pygmies, a stature estimate for the LB1 *floresiensis* skeleton came in at only 106 cm, or about 3 feet 4 inches tall.[334] All fauna that evolve on islands tend to be somewhat diminutive in stature, though some researchers have suggested that *Homo floresiensis* may have been exceptionally small because of a genetic abnormality that reached fixity within the population.[335 336 337]

Long-term height variability is also apparent in looking at African *Homo erectus*, which was among the tallest of our hominin ancestors. Even the earliest east African *H. erectus* specimens, dating as far back as 1.5–1.8 million years, ranged from 147–173 cm, or 4' 8" to 5' 7" in height.[338] In fact, based on a number of skeletal metrics and stature formulae, it is estimated that the Nariokotome boy, a juvenile *Homo erectus* specimen dating to between 1.51–1.56 million years ago,[339] would have reached an adult height of about 185 cm, or 6' 1".[340] Even compared to modern humans of the same age, which was about 11 years old based on his stage of dental development, the Nariokotome boy would still be in the 97th percentile for height when compared to a large reference population of New South Wales schoolchildren.[341][342] Beyond this one individual, the overall stature estimate for east African *Homo erectus* is about 160 cm for females and 180 cm for males, which is near the average female height of 161 cm and above the average male height of 175 cm, for adult modern humans.[343]

These examples of short and long-term changes in body height indicate just how variable this trait can be, and reported differences among extratempestrials suggest that further geographic, secular, and evolutionary height variation will also characterize the human future. Stature is just one of many features marked by a high level of variability, though many other traits have trended in the same general direction across the whole of human evolution. For instance, a reduction in body hair, increased finger dexterity, changes in pelvic and lower limb morphology, reduced tooth size, reduced and retracted faces, and changes in our overall brain size and shape, have followed the same general pattern throughout hominin evolution.

It is this very long pattern of continuous change, among these and other morphological characteristics, which helps support the notion that extratempestrials are simply us in the future. Additionally, modern peoples who have come into contact with these aliens of time provide matching—though occasionally variable—descriptions, which are all still consistent with what we would expect to see among individuals representing different stages of our distant human future. Based on past trends alone, it is difficult to predict when we may come to possess any of these traits of our earthly heirs. Though given the tenacity of these trends across vast swathes of hominin evolution, it is likely that they will continue to persist and eventually reach the point at which we will no longer need to look to the future for answers, as we may only need to look in a mirror.

10

Becoming Our Extratempestrial Descendants

Darwin's theory of evolution is a framework by which we understand the diversity of life on Earth. But there is no equation sitting there in Darwin's 'Origin of Species' that you apply and say, 'What is this species going to look like in 100 years or 1,000 years?' Biology isn't there yet with that kind of predictive precision.[344]

– Neil deGrasse Tyson

You have to know the past to understand the present.[345]

– Carl Sagan

10.1. Structural/Functional Tradeoffs in Hominin Evolution

10.1.1. The Obstetric Dilemma

In spite of minor variability in specific traits such as stature, skin tone, color and amount of hair, etc., extratempestrials are ubiquitously described as having outsized heads, large eyes, diminutive faces, and small noses and mouths. These characteristics represent an extension of the enduring trends of encephalization and facial orthognathism that have epitomized the whole of hominin evolution, which suggests that they are not likely to stop anytime soon. This may also be indicated by the fact that we have recently added cesarean sections to the modern human toolkit, which could help mitigate what has long been a barrier to further cerebral development during the earliest stages of our ontogeny.

Upright standing, walking, and running require a somewhat specific postcranial anatomical structure. Additionally, being eutherian mammals, females require a certain postcranial form that allows for gestating and bearing offspring. When our high level of encephalization is considered in this equation, it is easy to see how competition among structural and functional features of the cranium and pelvis results in a uniquely human quandary. This morphological rivalry between the human head and pelvis is known as the *obstetric dilemma*. More specifically, this refers to conflicting evolutionary pressures related to maintaining a postcranial structure that is conducive to bipedal locomotion, but also one that is capable of birthing infants with increasingly larger brains.

For decades, maintaining an efficient bipedal gait in hominin evolution was considered a limiting factor in the development of wider pelvises in females, which could otherwise circumvent what anthropologists refer to as the *big-head small-hole problem*. However, according to recent research, pelvic constraints associated with bipedal efficiency may not be as important to this complex anatomical equation as previously thought. For instance, using a biomechanical approach to examine gait efficiency and pelvic width, both within and between groups of males and females, researchers in the Department of Human Evolutionary Biology at Harvard University demonstrated that pelvic size was not significantly correlated with the efficiency of bipedal locomotion in humans.[346] According to the authors of this 2015 study:

> The results show that pelvic width does not predict hip abductor mechanics or locomotor cost in either women or men, and that women and men are equally efficient at both walking and running. Since a wider birth canal does not increase a woman's locomotor cost, and because selection for successful birthing must be strong, other factors affecting maternal pelvic and fetal size should be investigated in order to help explain the prevalence of birth complications caused by a neonate too large to fit through the birth canal.[346]

These researchers went on to list some of the other factors that could affect maternal pelvic and fetal size. These included evolutionary change in gestation lengths, rates of fetal ontogeny, mother's diet and fetal nutrition, thermoregulatory demands, the need for speed in bipedal running, pathogenic load, malnutrition, limits to maternal metabolism, and other environmental influences that are also important to consider in this anatomical relationship.[346][347] Additionally, another recent study by Betti and Manica (2018) showed that female pelvic canal shape is highly variable across geographic groups, which, rather than being the result of an evolutionary tradeoff between parturition and bipedal locomotion, was related to neutral evolution by means of genetic drift, differential migration patterns, and to a limited extent, climatic adaptation.[348] Together, this research indicates that selection for bipedal energy efficiency may not be a limiting factor for *in utero* brain development. However, because of the complexity of our evolutionary history and our morphological bauplan, there remains the vexing question of why we continue to suffer from this frequently fatal, fetal-head to maternal-birth canal size discrepancy.

Hominins did evolve some mechanisms that allowed us to continue growing larger brains, while still being able to get out of our mothers alive. For instance, we developed a fontanelle, or "soft spot" at the top of our skulls, which allows us to become somewhat cone-headed while exiting the womb. The downside of this configuration is that our growing brains are more exposed during our early formative years, which represents a quintessential example of these common structural/functional tradeoffs in hominin evolution.

In addition to the fontanelle, we also evolved extra cranial sutures, including the metopic suture that bisects the frontal bone in the sagittal plane, meaning that it runs down the center of our foreheads. This particular suture disappears in most people shortly after birth, once it has served its purpose of helping us squeeze through the narrow human birth canal. Though among the most notable evolutionary changes related to the big-head small-hole problem is our prolonged life history cycle, meaning that we are born in an underdeveloped state, and we develop more slowly relative to our hominin ancestors and extant ape cousins. This distinctive pattern of growth and development means that we are able to exit the womb earlier and with a comparatively smaller brain that, while underdeveloped, is still able to pass through the birth canal.

This elongated life history cycle is also germane to our extreme form of paedomorphosis and our high reliance on culture. For instance, being born in an underdeveloped state allows us to do most of our brain growth outside the dark and stimulus-deficient uterine environment. Therefore, as a species that relies heavily on culture, language, and learned behavior for survival, exiting the womb early allows us to grow our brains in a stimulus-rich environment, where and when we are able to absorb more of the culture around us. In this sense, the benefits of an extended and/or slowed rate of growth and development are twofold: 1) we are able to pass through the maternal pelvis before our brains grow so large that we become stuck and die and, 2) we are primed to learn important things about our environment and culture, at a time when our brains are growing the fastest.

The obstetric dilemma has caused a number of problems among big-brained hominins and in our recent evolutionary history. Not least among these is the possibility of both infant and maternal death, prior to, or during childbirth. In fact, it is estimated that as many as 20–25% of births resulted in death of the infant or mother during childbirth throughout human evolutionary history.[349] This seemingly high estimate of fetal/maternal death rates during parturition may actually be rather low as well. This is indicated by a late 19th century study of the Kuna Indians of Panama—conducted at a time when traditional medical practices were still used in deliveries—which found that 30% of mothers died while giving birth to their child, most commonly from obstructed labor.[350]

It is difficult to know exactly when this problem began to develop in hominin history, though from the innominate (pelvis) of our Australopithecine ancestor Lucy (AL 288-1), it is clear that our pelvic morphology had become quite modern by about 3.5 million years ago. At that time, our brains were still only about the size of a modern chimpanzee's brain, so we may have slid right out without any complications. However, by about 2 million years ago, our own genus *Homo* had developed larger brains and an even more modern postcranial skeleton, so we may have begun to push the boundaries of the birth canal a bit farther. The obstetric dilemma likely become even more problematic beginning 800,000 years ago, as hominin brain size began to increase at an accelerated rate, and then again around 200,000 years ago in anatomically modern *Homo sapiens*, as a result of increased cranial globularity.

Even with these obstetric constraints, encephalization and neurocranial globularity continued to march on. Moreover, whatever forces were driving the runaway brain train were clearly in control, especially considering that 20–25% of mothers died in childbirth, in spite of mitigating many of these obstetric constraints by means of a prolonged life history cycle, extra cranial sutures, thin and malleable cranial bones, and a fontanelle. Yet, this tremendous selective pressure—which killed mothers and offspring with an infant-head/maternal-pelvis mismatch—was still not enough to slow the unrelenting expansion of our brains. According to Weiner et al. (2008), in their article *Bipedalism and parturition: an evolutionary imperative for cesarean delivery?*:

...what is most remarkable in the evolutionary history of humans is the obvious strong selective pressure for increases in brain size. The obstetric dilemma has at its essential core the necessity for large brains as part of the package of emerging humanness. Selection was apparently so strong for the ensuing complex behaviors that came to characterize humans that it more than compensates for the difficulties associated with the birthing process.[351]

The introduction of cesarean sections in more recent history could certainly change this age-old antagonistic relationship between the human brain and birth canal. However, the degree to which it affects the morphogenesis of our crania and pelvises cannot yet be known, considering that we have only been performing C-sections for approximately 500 years.[352] In the context of the last 300,000+ years of difficult and deadly childbirth, this is but the blink of an eye. Though one would expect that this vital contribution to infant and maternal health may also act to relax what is arguably the last selective pressure keeping our brains from becoming too big, or too bulbous, too early. Additionally, the increased frequency with which this procedure is performed may indicate that we are entering a new phase of intensified in utero cerebral development. For instance, in looking at all U.S. births over a roughly 45-year period, the rate of cesarean sections went from 5% in 1970, to 20% in 1996, and all the way up to 32.2% in 2014.[353]

It is also important to acknowledge the economic incentive for those performing C-sections. There is ample evidence to suggest that this rate increase is partly the result of women being coaxed, or outright forced into having this procedure, even when it is not medically necessary. This is noteworthy, considering that obstetricians and hospitals earn far more money from this procedure as compared to a traditional vaginal birth.[354] [355] [356] However, other sociocultural factors could also be at play here, considering that affluent people in developed nations often opt for a "cesarean on maternal request." These procedures are most commonly performed without any indication of medical necessity, but are carried out due to fear of the severe pain of vaginal birth, fear of tearing during birth, later sexual problems resulting from vaginal delivery, or simply due to the added convenience that a scheduled cesarean section offers.[357] [358]

Pecuniary incentives and affluenza aside, there is no doubt that C-sections represent a monumental step forward in enhancing infant and maternal health, as this procedure has already saved countless lives when a baby is breech or unable to pass through the birth canal in other instances of obstructed labor. It is also important to consider the implications of this relaxed selective pressure on future human brain size, given that death was likely for both the mother and infant in instances of size/shape incompatibility throughout much of hominin history. Relieving humans of this longstanding evolutionary constraint on in utero cerebral growth would be expected to further brain size/shape changes at each stage of ontogeny, particularly if the driving forces behind encephalization and increased neurocranial globularity persist into the future. Weiner et al. (2008) also probed some potential implications of increased cesarean sections in the human future.

> How far will this trend go? What will be the consequences of universal cesarean delivery? Can we predict the possible effects on humans of almost total disassociation of childbirth from vaginal delivery…? Will the end result be a change in the overall balance between fetal brain size and the biomechanical properties of the pelvis associated with birth and with the limitations imposed by successful bipedal functionality? If yes, it is impossible to predict the direction or extent of that change.[351]

Indeed, it is currently impossible to predict the direction and extent of changes to infant and adult cerebral capacity based on the increased incidence of cesarean sections. However, if we can consider the cranial morphology of our presumed evolutionary descendants in the context of the role that cesarean section may play in allowing added cerebral development—given that if they are us in the future, then this is more of an observation than a prediction—it may be safe to say that the future direction and extent of this change is toward larger and rounder. Furthermore, if our brains are to someday attain the size and shape of our extratempestrial descendants, a complete disassociation of childbirth from vaginal delivery is likely, as cesarean section may become a medical necessity for all women in the distant human future.

10.1.2. Encephalization and Craniofacial Integration

Throughout the evolutionary history of our genus *Homo*, bigger was undoubtedly better, as our brains and neurocrania experienced a nearly threefold increase in size, in conjunction with increased social, linguistic, and cultural complexity. As mentioned previously, the role of intelligence and selection for greater cognitive abilities, problem solving skills, cunningness, and mate attraction were important to our unprecedented level of cerebral development. This is apparent in looking back through human history and prehistory, as well as from numerous intraspecific studies demonstrating a positive correlation between brain volume and intelligence.[359 360 361 362]

Recent research has also shown that human brains boast more overt levels of fluctuating asymmetry relative to chimpanzees, which indicates a high degree of developmental plasticity in our brains, meaning that they are capable of altering neural connections during growth and development in response to environmental interactions. This was undoubtedly a significant contributing factor to the intra- and inter-specific evolution of hominin culture and cognition.[363 364] Numerous neuroscience and brain imaging studies have also demonstrated marked cerebral plasticity in humans, in which the macroanatomy of our brains can be seen to increase in response to learning, meditation, and repetitive tasks in the specific cerebral areas associated with the task being performed. This shows that our brains are capable of growing beyond their genetically preprogramed size, in response to various environmental stimuli.[365 366 367]

There is no one factor that caused the runaway brain to continue running at an accelerating rate throughout the last 800,000 years of human evolution, though increased cerebral growth, reorganization, plasticity, and cognitive development were clearly important to our survival, as the hominin brain continued to eclipse other neighboring cranial and postcranial structural and functional traits. In fact, in the same way that a multitude of postcranial maladies are associated with our distinctive form of habitual bipedal locomotion—as per the perils of bipedalism described earlier—numerous craniofacial complications may also be considered a consequence of our immense brains. For instance, the massive human cerebrum has been implicated in the disruption of structural and functional features associated with mastication, olfaction, respiration, and vision.[368 369 370 371 372]

Organisms are not composed of distinct and unrelated anatomical features, but rather, they possess traits that operate in conjunction with one another.[373] As such, it is important to consider variation among spatially proximate structures of the skull and how long-term evolutionary trends may impact their utility in a modern context. Evolution does not simply work to create bigger and better things, but rather, the morphogenesis of every trait, across all organisms, exists as part of a patently give-and-take process. Consequently, it is important to understand the interchange between evolutionary biology and physiological function, especially in the context of tradeoffs associated with encephalization and the subsequent reorganization of neighboring hard and soft tissue components of the hominin skull.

In light of these anatomical concessions, it is easy to see how a bigger, more forward-projecting brain—as well as the ensuing competition for space that this imparts on adjacent features—can result in so many uniquely human problems. This includes such things as crowded and impacted dentition, reduced respiratory function, the unfortunate capacity to choke and die, the inability to smell as well as other mammals, a misalignment of our jaws with the mandibular fossa which can result in Temporomandibular Joint Disorders (TMJ), and, for an increasing number of individuals—particularly those who possess more neotenous craniofacial features—reduced vision that begins in early adolescence. While these inimitable traits of modern humans are rather unfavorable overall, they largely exist at the expense of the vast benefits imparted by big and intelligent brains. Additionally, in the same way that other morphological characteristics of past and modern humans may link us to extratempestrials in our evolutionary future, certain negative outcomes of this cerebral expansion might also help establish continuity through time.

10.1.3. Eye Size and Visual Acuity

Vision is an important sense among all primates. This is primarily the result of our long evolutionary history adapting to an arboreal environment and the associated need for keen vision and good depth perception, while moving about high up in the trees. However, throughout hominin evolution, the eye, surrounding ocular soft tissues, and the bony orbit around them, were sandwiched between an expanding brain and a retracting mid and lower face. This coalescence of craniofacial characteristics may have impacted our eyes in a functional sense, particularly during more recent stages of human evolution.

For instance, greater anterior and lateral development of the prefrontal cortex above the eyes, expansion of the temporal lobes behind them, and facial retraction below, would be expected to constrain orbital and ocular development among modern humans with this craniofacial configuration. These species-specific cranial traits—resulting from prominent trends of encephalization and facial orthognathism in hominin evolution—are even more important to consider in the context of different growth trajectories between the eye and brain, and the neurocranium and face.

Although the eyeball rests predominantly within the confines of the orbital walls, studies have repeatedly shown that the eye does not directly influence the size of the bony orbit in humans.[374][375] To the contrary, research examining ocular ontogeny indicates that the eye generally keeps pace with the brain throughout growth and development.[376][377][378] In fact, because the eye and the brain are so intricately linked, they are thought to be the product of pleiotropic gene control,[379][380] meaning that the same gene is responsible for guiding the development of both the eye and brain in humans.[381] This instance of genetic integration is certainly understandable as well, given that the optic vesicles—which mark the earliest stages of eye development—grow directly out of the forebrain during early fetal ontogeny.[382] Further evidence of this pleiotropic relationship is indicated by the eye and brain's common pattern of post-natal ontogeny, where both features grow at their fastest rate during early life, then cease growth at approximately the same age during late adolescence.

In the context of visual function and evolutionary tradeoffs with the hominin brain, if both ocular and cerebral growth are a product of pleiotropic gene control, accelerated and prolonged brain growth throughout hominin evolution would also be expected to increase the size of the eye. However, because of encephalization and brain globularity above the eye, and reduced facial development below it, increased ocular expansion may result in less available space for the eye and extraocular tissues that lie within the orbit. Additionally, and perhaps most importantly, while the brain is able to bend and fold within the confines of the cranium, the eye cannot.

As the eye grows larger in association with the expanding brain in hominin evolution—within the confines of an orbit that is unable to accommodate this added growth—one would expect our eyeballs to become a bit squashed. Additionally, because the eye is surrounded by bone on all but one side, the enlarged and constrained eye would become axially elongated (i.e., lengthen from front to back), simply because its only path for escape is forward, out of the front of the orbit. This is important, because the most common form of reduced visual acuity in humans is myopia, or nearsightedness, which is primarily the result of an overly large and axially elongated eye.[383][384][385]

Nearsightedness has been increasing in frequency and severity in human populations over the last 100 years. It also occurs at a disproportionately higher rate among more craniofacially paedomorphic groups. For instance, myopia affects 80–90% of individuals in many East Asian populations;[386][387] it is found to occur earlier in life and at a higher frequency among Chinese schoolchildren compared to African and European groups;[384 388] it occurs at a higher frequency among women in all ancestral groups; and on average, females develop myopia earlier in life, and have a higher refractive error (worse vision) once growth ceases.[389 390 391]

Throughout the majority of hominin evolution, the brain was located farther back behind the eyes and the face was more elongated, which kept the brain, eyes, and face broadly separate from one another relative to anatomically modern humans. Additionally, natural selection would have acted to maintain high visual acuity throughout much of our pre-history, due to the important role that vision played in the day-to-day activities of hunter-gatherer groups. However, during more recent stages of human evolution—following the advent of agriculture in particular—selection for acute vision may have been relaxed due to changes in subsistence strategy, social and political organization, new forms of economic exchange, and in association with human self-domestication.[372 392] Furthermore, during even more recent stages of human biocultural evolution, we have become capable of mitigating the effects of poor vision with glasses, contact lenses, and refractive eye surgery.

These mitigating mechanisms arose in conjunction with our increasing intellect and advancing culture, which, somewhat ironically, are products of the same brain that is implicated in these and other evolutionary tradeoffs. For instance, we can alleviate pain from TMJ with surgery, we have masks and machines to provide continuous positive airway pressure to help with sleep apnea, we can change the refraction and focal point of light with glasses and contacts, and we have the ability to add and remove teeth as we see fit.

This ability to ameliorate anatomical tradeoffs associated with our unique form of biocultural evolution is an important component of the hominin past, present, and likely the future as well. Cesarean sections, glasses, contact lenses, the Heimlich maneuver, wisdom tooth extractions, etc., represent just some of the many cultural solutions we have developed in response to biophysical problems resulting from our big brains and advanced intellect. Furthermore, if we continue to reap the benefits of a larger, more complex, and better-organized brain—while maintaining the ability to minimize its negative consequences using cultural mitigation mechanisms—then what is to stop the runaway brain from continuing to run even faster.

In the quote that begins this chapter, American astrophysicist Neil deGrasse Tyson points out that we are not yet to the point where biology is able to precisely predict what our species will look like in the next 100,

1,000, or 100,000 years.[393] Based on past anatomical trends alone, this is certainly true. We cannot definitively say whether, or how much, our brains will continue to enlarge, or what effects this may have on adjacent characteristics in shaping our overall appearance. Though if we are afforded glimpses of our future form when our more advanced progeny trek from the future to our own time, this aids in removing certain predictive components of this hypothesis, as these intertemporal ties present to us the end result of what lies between now and then.

10.1.4. Extratempestrial Autopsy

In addition to pictures, eyewitness accounts, and the remains of a disc-shaped craft, the bodies of individuals inside that craft were also allegedly recovered from the crash site outside Roswell, New Mexico, in 1947. Moreover, an autopsy was allegedly performed on one of the crash victims, which was captured on film and later leaked by a military cameraman, who claimed to have been present during the examination. Questions relating to the authenticity of this now-prolific autopsy video will be discussed shortly, though it is referenced here primarily because of one specific and seemingly insignificant scene.

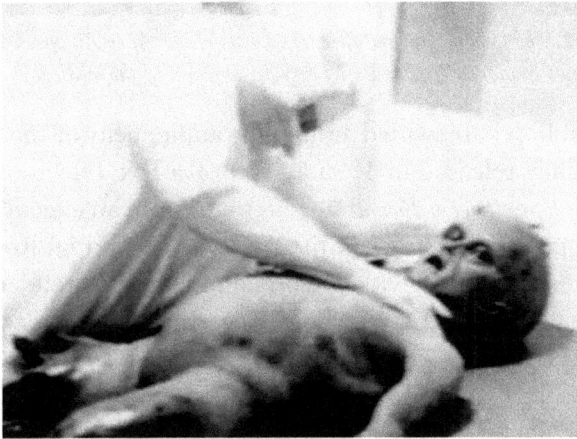

Figure 10-1. *Image from an autopsy performed on an extratempestrial/replica allegedly recovered from the IFO crash site at Roswell, New Mexico, in 1947.*

The scene in question involves one of the medical examiners removing darkened contact lenses from the eyes of an extratempestrial, which are in every respect overly large, yet overtly human eyes (figure 10-1).[394] This scene is important to the current discussion because, if these lenses are to help correct for vision defects—the same way they do in modern humans—it may represent yet another example of a species-specific structural-functional tradeoff in hominin evolution, which also helps connect our past, present, and future. Beyond this potential biocultural connection through time, on an even more basal level, this simple act of removing translucent contact lenses from the eyes of our deceased descendant helps reveal the distinctive human identity of this being. In other words, before these lenses are removed, the extratempestrial appears to have cold, dark, expressionless, and altogether alien-looking eyes. However, after the lenses are taken off, this individual suddenly appears much more human, with ocular anatomy that—while much larger—is identical to our own in every way (figure 10-2).[394]

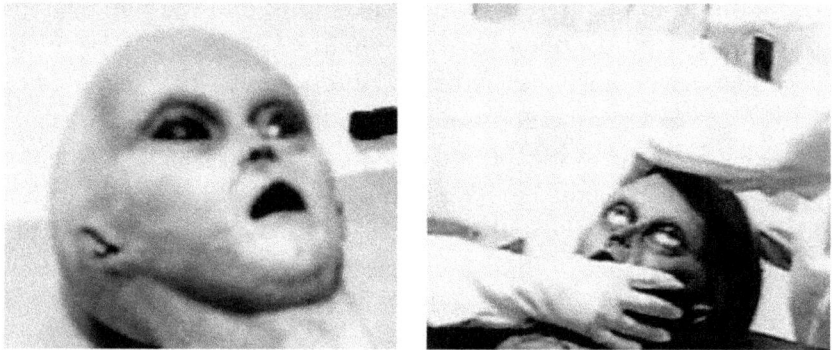

Figure 10-2. *Showing before and after screenshots of the Alien Autopsy video, with darkened contact lenses in place (left), and after they are removed by the medical examiner (right).*

Considerable debate swirled around the authenticity of this video when it was originally released in 1995 as part of a Fox Television broadcast titled *Alien Autopsy: Fact or Fiction*.[394] This stems from an effective campaign of instilling public doubt about the reality of the initial Roswell crash, the unusual nature of the video's content, as well as broader skepticism about the reality of aliens as a whole. Nevertheless, the video is indeed a fake. However, there is good reason to believe that it was based on—and a concerted effort was put into accurately recreating—real footage shot during an actual autopsy involving an extratempestrial that was indeed recovered from the IFO that crashed near Roswell back in 1947.

After reviewing the research of others who have examined this footage and its somewhat atypical history, the conclusion reached was that a videographed autopsy took place following the recovery of extratempestrial remains from the Roswell crash. The film was later leaked by a retired military camera operator, it was refurbished, the scene was recreated, and the *Alien Autopsy* film was released to the public as a sensationalized mainstream reconstruction of the initial event. This inference is further corroborated by Ray Santilli, one of the producers of the 1995 Alien Autopsy film.

In a 2006 interview on the British program *Eamonn Investigates: Alien Autopsy*, Santilli states that this was in fact a staged production. However, he claims that it was based on actual footage of an actual alien autopsy. After nearly 50 years in subpar storage conditions, the film was too badly deteriorated to use in its entirety, which is what prompted the theatrical recreation of this original event.[395] Much has also been written regarding the authenticity of the original 1947 autopsy film that inspired this reproduction. Yet, sitting above the fray, remains an issue that has not yet been considered, given that close encounters have thus far been viewed in the context of contact with extraterrestrial aliens, as opposed to interactions with future members of the hominin lineage.

If the individual in this autopsy video is regarded as a product of continued human evolution, rather than the result of a separate evolutionary trajectory on a separate planet, then the authenticity of this initial event, and the film that followed, become somewhat more palpable. For instance, if someone were trying to create a fake autopsy video that was not based on actual footage captured as part of an actual alien autopsy, then there is no reason why they would make the alien look so human. In fact, they would probably not even want to make it look human at all, considering that the prevailing paradigm involves an extraterrestrial explanation for the existence of aliens.

People attempting to make a fake autopsy video would likely not add so many uniquely hominin features to the alien—including specific human cultural implements like contact lenses—if they had not been informed by actual footage from an actual autopsy. With this in mind, the being in the *Alien Autopsy: Fact or Fiction* video was given an incredibly realistic human physiology, with two arms and legs, hands and feet with 10 fingers and toes that each had nails and tactile pads, and large eyes that were anatomically identical to our own. Additionally, and perhaps most notably, it possessed a chin, mouth, nose, ears, and neurocranial configuration that were allometrically proportioned in the same way we would expect to see them in our paedomorphic descendants.

If this reconstructed autopsy video was not derived from real footage of a postmortem examination—carried out on an actual extratempestrial recovered during a botched mission to 1947 New Mexico—one would think that the film's producers and creature artists would have made the extraterrestrial alien look much more like an extraterrestrial alien, rather than a slightly bigger-brained human with bad vision. These traits, as well as numerous other subtle human characteristics and unique derived features of our hominin clade, suggest that this recreated video autopsy was indeed heavily influenced by actual events, which occurred during a real medical examination involving one of our time traveling descendants.

10.1.5. Eyes and Brains of the Human Future

In addition to the black contact lenses on the normal human eyes of the extratempestrial in this autopsy video, this same dark-eyed trait can be seen on the extratempestrial on the cover of the aforementioned book *Communion*, by Whitley Strieber. As in the autopsy video, the black eyes of this individual also detract from its humanness, as it retains a far more "otherworldly" appearance, with cold, dark eyes. The apparent ubiquity of tinted contact lenses worn by future members of our species may also help explain the dark eyes of the extratempestrials described in the abduction reports detailed in the previous chapter. This includes one report that states, "I opened my eyes and could see nothing but black, very deep and inky black. As I was looking at this, the blackness pulled back, revealing a face, I had been looking into the eyes of a grey type creature."[396]

These potential intertemporal linkages could help inform other aspects of our evolutionary future as well, such as how the recent and mounting epidemic of reduced visual acuity in modern humans may play out in years to come. For instance, humans appear to have recently reached an important threshold at which acute eyesight can be traded for further neocortical development. The bigger eyes, as well as the larger and more forward-projecting neurocranium of our presumed distant descendants, suggest that these structural-functional constraints will persist, and worsen, long into the future. From the assemblage of abduction cases presented in the previous chapter, extratempestrials who were not wearing tinted lenses were described as having brown or blue eyes, which were always said to be larger and often more oval-shaped than our own.

Of course, it is entirely possible that these lenses serve the purpose of shading the sun, given that we too put dark things in front of our eyes for this same reason. It is also possible that the lenses are meant to help conceal their extratempestrial identity, given how well the lenses change their overall appearance, and also the great effort they put into remaining incognito while visiting past periods of time. Though given the mounting conflict between our brains, eyes, and faces as they vie for space upon the island of the skull, these are likely not the only function of tinted lenses in the distant human future. In fact, this aspect of reduced visual acuity among future humans may also be indicated by how often extratempestrials are likened to individuals of East-Asian descent, who also happen to have the highest rates of juvenile-onset myopia among any modern human group.

Extratempestrials are often compared to those with "Chinese" or "Asian" ancestry, and particularly due to their eye shape, where they are commonly described as having large and slanted "Oriental-looking eyes." For instance, Strieber described them as being slanted "more than an Oriental's eyes,"[397] and among the other accounts provided in the previous chapter, one abductee was quoted as saying they "looked somewhat Asian, with straight dark shoulder-length hair and dark eyes,"[398] while another individual—who apparently doesn't get out much—stated that "Once I saw Chinese men and they [the beings] slightly reminded me of them."[399]

IFO reports commonly describe extratempestrials as Asian-looking, which suggests that Asian researchers are destined to someday lead the way in studies involving backward time travel, assuming of course that geographic races still exist in the deep future. This commonly stated ancestral similarity could also indicate that our entire species is collectively moving toward this same craniofacial form. This assumption is supported by the more neotenous characteristics of East Asians, as well as our species-wide trend toward greater paedomorphism as a whole. Additionally, given that myopia rates are far higher in East Asians and females within every human population, and because both groups are highly neotenous, the argument could be made that something about a paedomorphic craniofacial form makes us more prone to developing nearsightedness.

In fact, this same relationship between craniofacial paedomorphism and refractive error is also observable among certain dog breeds. For instance, reduced vision occurs at a higher frequency among the Miniature Schnauzer, Pug, Rottweiler, Labrador retriever, and Toy Poodle,[400] [401] which also tend to be more craniofacially paedomorphic relative to other varieties. Additionally, myopia is less prevalent in breeds like the collie (35.7%), which are characterized by rate/time hypermorphosis, or ontogenetic trajectories more like the ancestral wolf, compared to the much more paedomorphic toy poodle, for example (63.9%).[402]

Domesticated dog breeds with worse vision are most often those characterized by greater craniofacial paedomorphism, resulting from artificial selection for the retention of juvenilized traits, and the conservation of "cute" into adulthood. Additionally, from descriptions of extratempestrials in reports of close encounters, the "alien" portrayed in the above autopsy video, and in association with persistent trends toward increased neoteny in human evolution, it would seem that we—much like the dogs we domesticated—will also retain more juvenilized traits throughout the hominin future.

10.2. Sustained Trends in Hominin Neoteny

10.2.1. Kindchenschema

As stated previously, the reported cranial and postcranial bauplan of ex-tratempestrials is completely in line with that of humans. It is also exactly what we would expect to see with the continuation of anatomical changes that epitomize the hominin past, particularly an ascendant trend toward greater neoteny in human evolution. This aspect of hominin morpho-logical change is associated with the retention of youthful characteristics into adulthood, which is an aspect of both our biology and culture. For instance, our species-wide online obsession with videos of kittens, puppies, and babies shows we have a strong affinity for youthful cuteness, which is actually a characteristic of most mammals that give birth to altricial young—meaning ones born in a relatively underdeveloped and helpless state.

This relationship between paedomorphosis and altriciality was first pointed out by Austrian zoologist Konrad Lorenz over 70 years ago. He hypothesized that the cute characteristics of our helpless young elicits a desire to nurture and provide resources for them, while suppressing feelings of frustration and aggression that often accompany the act of rearing difficult and demanding offspring.[403] In essence, he concluded that our altricial young look cute to encourage us to give them food and hugs, as opposed to giving them to leopards. This propensity to feel affection toward paedomorphic things isn't just a characteristic of humans either. In fact, a recent experimental study demonstrated that this preference for big headed, big eyed, small faced, paedomorphic traits extends well beyond our own species. More specifically, this research showed that increased neotenous cuteness—or Kindchenschema as Lorenz called it—elicits a powerful effect in humans and other animals that give birth to altricial young:[404]

> Our most interesting finding is that the effects of facial cuteness adaptation transfer across species . . . we found similar after-effects for human infant faces after participants were exposed to cute and less cute faces of puppy dogs. These findings support Konrad Lorenz'

(1943)[407] claim of a species-unspecific Kindchenschema. According to Lorenz, a cute infant face is determined by specific pedomorphic characteristics which are present in many different species (e.g., dogs, tigers etc.) ...This result can be taken as experimental evidence for the universality of the Kindchenschema.[408]

Numerous cerebral imaging studies have also shown that baby faces —and particularly our own baby's faces—trigger dopamine rewards in the brain, which subsequently leads to increased caregiver behaviors. Additionally, cuter babies provide us with higher doses of dopamine, resulting in even greater caregiver effort.[405] In a 2015 Wall Street Journal article titled *Why We Melt at Puppy Pictures*, author Robert Sapolsky states:

> Mammalian babies all look the same: short snouts, high foreheads, round faces, big eyes. And we love it. The more a particular baby's features accentuate those traits, the cuter we rate the face. People, including children, prefer pictures of babies over those of adults, and they prefer cute babies over un-cute ones.[406]

Interestingly, we apply these same basic evolutionary principles to fictional characters in an attempt to make them more appealing to both children and adults. This technique has even been used to make aliens appear less threatening on a number of different occasions. For example, films like *ET*, and especially *Close Encounters of the Third Kind*, depict the aliens with extremely paedomorphic characteristics, which make them look less menacing and even somewhat amiable.[407][408] In fact, the quintessentially large head, soft wide eyes, and general juvenile appearance and behavior of ET was enough to garner the affection of small children, acquire free Reese's Pieces, and easily blend in with a closet full of equally paedomorphic stuffed animals.

A recent pattern of juvenilization among animated characters also demonstrates the continued and increasing importance of Kindchenschema as an emergent cultural value. This conspicuous trend toward cartoon craniofacial paedomorphosis was first noted by renowned evolutionary biologist Stephen Jay Gould, in a 1979 article titled *Mickey Mouse meets Lorenz Konrad*. In this innovative and illuminating study, Gould documented the evolution of Mickey Mouse's craniofacial morphology, since the earliest incarnation of this famous cartoon character. As part of this study, Gould discovered a steady shift toward overtly accentuated paedomorphic traits across the 50-year history of Mickey Mouse.[409]

This trend certainly didn't stop with Mickey Mouse in 1979 either, as the neotenization rate for both human and non-human animated characters has continued to hasten. This is most apparent in looking back through the last 20 years of animation, where a marked shift toward neotenous features can clearly be seen, particularly among young, female, princess types (see figure 10-3). Changes to the bauplan of animated characters throughout this time are most discernable in how much the head has increased in proportion to the rest of the body and how the eyes and neurocranium have grown taller and wider, while the ears, nose, and mouth—if perceptible at all—have diminished substantially. Together, these create the cuter, more youthful, and more marketable physical appearance of both female and male cartoon characters of the recent past.

Figure 10-3. *Common female cartoon form exhibiting overt craniofacial paedomorphosis.*

To illustrate the extent to which we have over-accentuated these paedomorphic traits in recent times, an analysis of craniofacial change was carried out using a large sample of 230 human animated characters, taken from different North American television programs spanning an 84-year period. The temporal range for this sample begins with Betty Boop and Snow White in 1932 and 1937, respectively, and ends with the most recent Disney Princess Moana Waialiki, from the film *Moana*, which was released in late 2016. Additionally, it should be noted that to avoid sample bias, all cartoon characters used in this analysis were selected by a research assistant who was blind to the study objectives.

Borrowing from Gould's 1979 paper, digital calipers were used to measure and calculate eye/head, eye/face, and face/head height ratios for each individual in the sample. Three separate regression analyses were then performed to test whether these craniofacial ratios changed over time, which was determined by the year each program first aired on television. This regression analysis was also carried out to assess whether a shift toward more paedomorphic proportions occurred among human cartoon characters, as it had in Gould's prior study of Mickey Mouse.

The results of these analyses showed that each of the above ratios were significantly correlated with time, and that the direction of this change was consistent with a patent shift toward more neotenous characteristics. For example, eye height relative to overall head height was found to be positively correlated with the original air-date of these programs (F(229) = 36.5, p < 0.001). In fact, eye size increased approximately 19% over this 84-year period—or about 0.22% per year (figure 10-4). This dominant trend in the relative eye height of these human cartoon characters exemplifies the increased importance of paedomorphism and Kindchenschema as a cultural value in North American society, given that proportionately larger eyes are among the cutest characteristics of human and nonhuman animals alike.

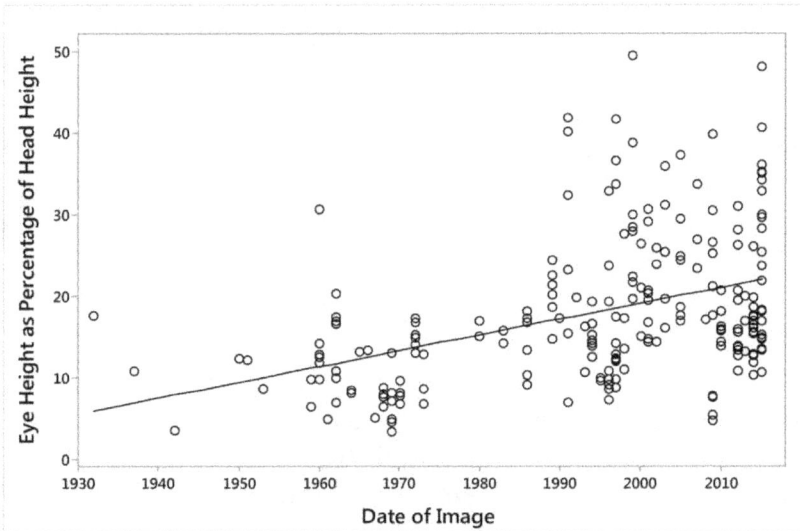

Figure 10-4. *Scatterplot with regression line showing a notable change in the eye/head ratio of cartoon characters across an 84-year timespan, signifying the increased cultural value placed on the retention of juvenilized traits since 1932, most notably since the 1990s.*

The increased cultural importance of a more youthful appearance can be inferred from figure 10-4, where the biggest upsurge in eye/head height occurred only in the last few decades—as indicated by the large cluster of human characters that sit above the regression line beginning about 25 years ago. This represents a rather dramatic shift away from the slow but steady increase that had already been occurring since the 1930s—at a time when most human animated characters had eyes that were only about 8–10% of their total head height. Although, beginning in the early 1990s, some started to be drawn with eyes that took up a larger share of their heads, even approaching 50% of total head height at times. Assessing patterns of change in the features of iconic pop-culture characters can help gauge epitomes of paedomorphic cuteness and how the idealized male and female form changes over time. This is also important because such perceptions and portrayals both mirror and help to create cultural values tied to human appearances, within and across geographic and temporal groups.

10.2.2. The Yin and Yang of Culture and Biology

The recent tendency to neotenize the heads and bodies of human and non-human cartoon characters indicates that we put a high value on these cross-cultural and cross-species ideals of cuteness. These changes to our animated characters are certainly not the cause of human neoteny, though they do reflect and, on some level, may help shape sexual preferences and reproductive behaviors, which could contribute to directional sexual selection in some form. After all, Betty Boop—arguably the first to possess these overly accentuated neotenous traits—quickly became an international sex symbol and cultural cartoon icon that is still widely revered today, despite starting out as an anthropomorphic female dog in the 1930s cartoon *Dizzy Dishes*.

Betty Boop's colossal eyes and tiny face on a scantily clad voluptuous body has been plastered across various forms of pop-culture media over the last 84 years, which certainly could have helped shape perceptions of iconic beauty through time. As this cartoon phenotype gained a larger media presence, it is reasonable to assume that these socially constructed ideals could have occasionally manifested themselves in the actual mate selection behaviors of real living people. Though even beyond any individual effects, this foremost trend toward increasingly big-eyed, big-headed, small-faced cartoon characters—among both sexes and across all age cohorts—clearly speaks to the increased importance of Kindchenschema in human culture and possibly human biology as well.

Looking across the entire study sample, Betty Boop's rather egregious 1932 craniofacial form is actually now toward the bottom end of the eye/head spectrum for modern-day cartoon characters, who tend to cluster well above her 17.6% value for this ratio (figure 10-4). This practice of over accentuating neotenous characteristics in animated characters has also been a part of Japanese culture for nearly 60 years. In fact, the origins and spread of this popular Japanese style of animation, known as *Anime*, may also help frame the significance of this synergistic relationship among cultural ideals of beauty, the way in which a society depicts animated human characters, and variation among modern human populations with regard to the incongruent phenotypic expression of paedomorphic traits.

Also known as Japanimation, or Japanime, this animation style originated with legendary Japanese manga artist Osamu Tezuka in the 1940s and grew in popularity throughout the 1960s and 1970s. Tezuka was heavily influenced by the disproportionately large eyes and childlike features of Betty Boop and Snow White in the 1930s,[410] which helps explain the paedomorphic characteristics, and especially the large eyes, small faces, and childlike bodies of the Anime characters that followed. However, while this juvenilized form would persist and even go on to define the Japanese style of animation, it was almost fifty years before Western animators returned to depicting their characters with the same neotenous traits that first inspired Tezuka during the 1940s.

Considering this temporary inter-cultural discrepancy, one can't help but wonder if the early appeal and persistence of these juvenilized traits in Japan were somehow related to the fact that the Japanese themselves are among the most paedomorphic of any modern human population. Their early enthusiasm for these physically and behaviorally paedomorphic cartoon characters—which have only increased their suite of juvenilized traits since that time—likely reflects, and may have even helped to perpetuate, neotenous change within this society.

By contrast, Western animated characters of both sexes were generally depicted with smaller, deep-set eyes and with rugged, highly masculine characteristics. To some extent, this also mirrored the craniofacial phenotype of actual adults in Western culture at that time. This East-West dichotomy reflects the divergent embodiment of an ideal physical form between different regional societies in the recent past. However, a more contemporary convergence of these animation styles would seem to suggest that they are now trending in the same direction and at a more congruent rate. This is also evinced by the extent to which individuals of European ancestry have changed in craniofacial form within the much more recent past.

For instance, it is clear from looking at European-descended women in old pictures, paintings, and movies—even as recently as two to three generations ago—that they possess a much more masculine craniofacial architecture, at least compared to the full-lipped, small-chinned, small brow, and round face that is so common among women today.[411] Additionally, males a few generations back can be seen to have more hypermasculine traits relative to modern males within this same ancestral group, which is a pattern of change that is tied to both our culture and biology.[412] A preference for feminized male and female faces—which acts to encourage neoteny and limit sexual dimorphism in humans—is actually part of a notable trend toward craniofacial feminization in humans that has persisted irrespective of changing hair styles, modes of dress, and other cultural variables. This is certainly not a new trend either, as researchers in various academic fields have consistently demonstrated this prominent shift toward male feminization, and female ultrafeminization, over the last 80,000 years of human evolution.

10.3. Self-Domestication and Craniofacial Feminization

Long-term trends of encephalization, reduced facial prognathism, and increased craniofacial neotenization, are defining features of the hominin lineage. Considering the even larger and more bulbous brains, larger eyes, and smaller faces of our reputed extratempestrial descendants, these dominant patterns of morphological change are likely to continue long into the future, along with the enduring forces that helped shape them through time. Many factors have guided this directional shift throughout our evolutionary history, surely with an ebb and flow in the degree of influence of any one component at any given time. Though more recently, this accelerating trend toward the further retention of juvenilized physical and behavioral characteristics into adulthood appears to be linked to a more recent process of self-domestication, which began around 8,000–10,000 years ago.

Humans are clearly the main driving force behind the domestication of dogs, cows, sheep, and countless other animals, meaning that they have been altered over many generations to be more docile and to live in close association with humans. However, many people don't realize that we are also responsible for domesticating ourselves. We can easily observe in humans many of the same traits that arose in other animals after we domesticated them. In fact, researchers recently met at the Salk Institute for Biological Studies in California to discuss these anatomical and behavioral traits shared among domesticated human and non-human animals, at what was the first-ever symposium on human self-domestication.[413]

Among the most important factors associated with human self-domestication was the development of larger and more complex social groups.

Following the origin and spread of agriculture and the rise of civilizations, individuals were forced to get along with one another, as they came to rely on social networks to a greater extent than in the past. In this sense, we began to trend toward prosocial behaviors, cooperative communication, and an overall tameness toward our fellow humans.[414][415] Over time, more intelligent, cooperative, and politically motivated individuals were selected over vicious warmongering types, who may have been preferable in earlier, less civilized societies, with different forms of social, political, and economic organization.

Others at the symposium focused on the shift toward, and underlying processes behind, the trend toward more feminized paedomorphic traits of the human skull, which has accelerated among modern humans over the last 80,000 years. Numerous explanations have been put forth to explain why craniofacial feminization has been occurring to such an extent in recent human evolution. One such theory ascribes these morphological changes to lower testosterone levels, driven by natural selection, sexual selection, and social selection for increased tolerance.[416] Others point to a more basal, yet somewhat more holistic mechanism, involving developmental changes in embryonic neural crest cells, which are stem cells that grow out of the dorsal section of the neural tube in vertebrate embryos.[417]

In a 2014 paper, researchers from the University of Vienna, Harvard University, and Humboldt University in Berlin, Germany, provided evidence for what they consider to be the first unified theory of domestication. More specifically, this research examined the mechanisms responsible for the *domestication syndrome* (DS), defined as the entire suite of phenotypic characteristics—including increased docility, reduced pigmentation, more frequent and nonseasonal estrus cycles, reductions in tooth size, perpetuation of juvenile physical and behavioral traits, etc.—which are observable among domesticated human and non-human animals:

> In a nutshell, we suggest that initial selection for tameness leads to reduction of neural-crest-derived tissues of behavioral relevance, via multiple preexisting genetic variants that affect neural crest cell numbers at the final sites, and that this neural crest hypofunction produces, as an unselected byproduct, the morphological changes in pigmentation, jaws, teeth, ears, etc. exhibited in the DS.[421]

In other words, these researchers argue that the suite of traits associated with tameness and domestication can be traced to reduced numbers of, reduced migratory ability of, and/or diminished proliferation of neural crest cells, to the sites where they would otherwise help to create the wild-type condition in undomesticated versions of these same species. This unified theory helps explain the underlying mechanisms responsible for driving the accelerated rate of self-domesticated paedomorphosis in our species, as well as this tendency toward craniofacial feminization throughout the recent human past. In fact, when examining variation in male and female skull shape on a scale from hypermasculine to hyperfeminine, a number of notable patterns emerge, which, taken loosely, may help inform the craniofacial shape of future humans, if we are indeed moving toward becoming our paedomorphic extratempestrial descendants.

10.3.1. Allometry

One of the best ways to visualize craniofacial shape differences between the sexes is with geometric morphometrics, which is a commonly used tool in biological anthropology research. Geometric morphometric methods use Cartesian geometric coordinates—as opposed to linear, area, or volumetric variables—to study morphological variation associated with evolutionary, ontological, and other biological processes. This method allows researchers to separate size from shape and to quantify and visualize variation in the physical form of organisms. These analyses, both with and without size as a factor, provide different but complimentary information that can be applied across a wide variety of scientific disciplines.

An early pioneer in this field of mathematical biology was Sir D'Arcy Wentworth Thompson, whose seminal 1917 book, *On Growth and Form*, became central to the biological sciences. His research was particularly impactful regarding our understanding of patterns of ontogeny, plant and animal morphology, the macro-form of organisms, and with regard to allometry, meaning the change in an organisms body proportions as their various parts grow at different rates in relation to one another.[418] More recently, with the development of novel quantitative methods for statistical shape analysis, as well as the advent of new and more powerful computer technologies, landmark-based geometric morphometrics has become one of the most prolific tools for examining phenotypic variation in scientific research.[419]

In order to better understand morphological changes associated with craniofacial feminization—and to visualize potential future human skull shapes if this and other hominin trends were to continue—a geometric morphometric analysis was carried out as part of the current study. More specifically, average craniofacial shape differences between female and male skulls were examined using standard anatomical landmarks and the European Virtual Anthropology Network (EVAN) Toolbox software, version 1.70 (figure 10-5).[420]

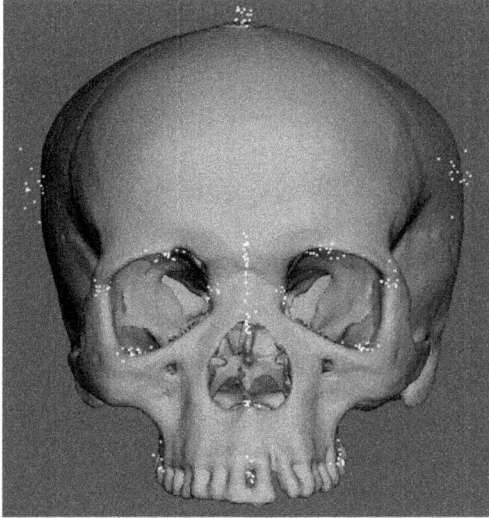

Figure 10-5. *Standard anatomical landmarks used in this geometric morphometric analysis.*

Once size is factored out—most notably sexual dimorphism, meaning size differences between males and females—overall human craniofacial shape variation becomes much more apparent. This generalized shape variability is particularly evident in moving away from the average genderless male/female form at the center of figure 10-6. Here it can be seen that the extreme male shape is characterized by a narrower and more elongated skull, a projecting face, and a lower and flatter forehead. In contrast to this accentuated male shape, the extreme female shape is exemplified by a much shorter face, a taller and wider forehead, and a generally more bulbous neurocranium by comparison.

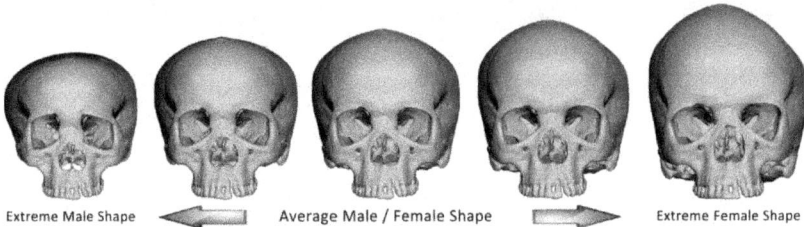

Extreme Male Shape ⟵ Average Male / Female Shape ⟶ Extreme Female Shape

Figure 10-6. *Shape variation between hypermasculine (left) and hyperfeminine (right) craniofacial forms*

Warping the image in the 3D Cartesian coordinate system toward the extreme ends of this sex-specific shape spectrum mirrors long-term patterns of craniofacial shape change throughout hominin evolution. This morphological pattern is apparent in moving from the ultra-male form (left side of image) toward the average of males and females (center) and on toward the ultra-female craniofacial form at the right-hand side of the image. More specifically, the hypermasculine shape to the left is somewhat indicative of our earlier *Homo* or Australopithecine ancestors where it can be seen to have a lower, narrower, and anteroposteriorly elongated neurocranium, a shorter forehead, greater post-orbital constriction—meaning a narrower cranial breadth behind the eye orbits—as well as a much more projecting face.

These long-term evolutionary changes—reflected here in the context of extreme shape differences between the sexes—are even more recognizable when looking at the profile view of these same skulls (figure 10-7). In this lateral view, warping the 3D Cartesian coordinates toward the extreme male end on the left-hand side of the image reveals an increasingly more primitive shape, with a low forehead and highly prognathic face, which is much more representative of our earlier hominin ancestors. In this same way, moving toward the extreme female form on the right-hand side of figure 10-7 may offer some insight into the generalized craniofacial shape of our future human descendants. This is particularly noteworthy given the persistence of long-term craniofacial trends in hominin evolution as they relate to encephalization, facial reduction and retraction, neoteny, and an equally prominent shift toward feminization of the modern human skull during more recent stages of human evolution.

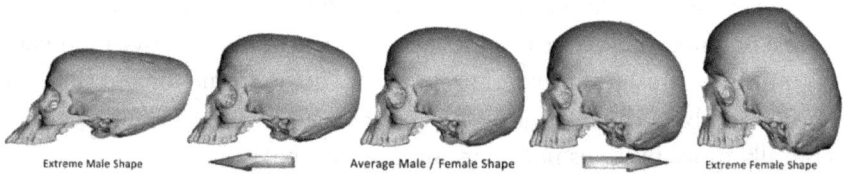

Extreme Male Shape   Average Male / Female Shape   Extreme Female Shape

Figure 10-7. *Hypermasculine (left) and hyperfeminine (right) craniofacial shape variation*

It is important to emphasize that this sex-based morphometric warping exercise should not be taken as an assertion that human craniofacial morphology is someday destined to resemble the hyperfeminized skull at the right-hand side of these images. Rather, this is simply an indication of a form that is likely, based on sex-specific anthropometric variation among modern humans, long-term trends of encephalization, reduced facial prognathism, and paedomorphosis in the hominin lineage, as well as the continuation of a prevalent trend toward craniofacial feminization, which has occurred in association with self-domestication over the last 80,000 years of human evolution.

Nonetheless, it is hard to ignore the extent to which this ultra-feminine craniofacial shape resembles that of the individual depicted in the above autopsy video and in descriptions of extratempestrials from reports of close encounters. Although we cannot currently predict what our skull shape will look like in the distant human future, persistent trends toward bigger and more globular neurocrania, smaller more retracted faces, a universal appetite for the Kindchenschema, and a prolific feminization and neotenization of our cranial and facial morphology, are likely to persist for even longer. Furthermore, if we can also take into consideration these other more uncertain forms of extratempestrial evidence in the form of close encounters and similar other sources, the suggestion is that our species will continue to move toward a generally more feminine, youthful, and globular craniofacial shape in the human evolutionary future.

10.4. Becoming Our Extratempestrial Descendants

While discussing this intertemporal hypothesis espousing linkages among hominins through deep time, I am commonly asked: *"when* might we come to look like these extratempestrials visiting us from the future?" Unfortunately, the response is often somewhat disappointing. This is because it is problematic to plot an organism's rate of evolutionary change even across the much more tangible past. So naturally, making predictions about changes to the future form of an organism is even more challenging. This is especially true regarding the future of human evolution, given the role of culture as both a product of and a contributing factor to our biological evolution.

Initially, the intent was to draw a sort of morphological timeline that would plot prior changes in size and shape of the hominin craniofacial skeleton and project them into the future. It was thought that such a timeline might help those who have had an extratempestrial encounter to be able to pinpoint the future time, or times, from which these individuals came. However, attempting to determine the home time of our extratempestrial descendants, based upon where they lie along an extrapolated future human morphological continuum, is more akin to erratic conjecture than concrete calculation.

Among the most problematic aspects of this type of prediction is our acceleratory rate of morphological and cultural change, particularly during more recent stages of human evolution. Other difficulties center on the tremendous number of biocultural factors at play—and the relatively short period of time that many of these have been occurring—that make extrapolating them into the future an even thornier affair. Because changes to human craniofacial morphology have been accelerating, and because numerous other cultural and environmental factors have entered the fray more recently, it becomes incredibly difficult to make sound predictions about the timing of specific morphological aspects of the human future.

As described earlier, there is also the confounding issue of sex, age, and population-specific variation as we move farther out away from our current position in time. This is an obvious limitation associated with looking into the future without the aid of a time machine, as events beyond the present remain blanketed by the veil of time. Chronological sampling bias is particularly problematic regarding the temporal origination point of extratempestrials. This is because we have no way of dating them, or gaining insight into how much of their variation is due to sex, age, geographic, or evolutionary variation. Especially considering that different extratempestrials, observed during any one encounter in our own time, may have originated from different points in our evolutionary future.

The desire to speculate about what factors will contribute to our evolutionary future is certainly strong. This is apparent from looking at any number of science fiction books, television programs, and movies. Though given the intense synergistic relationship between biology and culture throughout hominin evolution—as well as accelerating rates of change in both culture and biology in the recent human past—any overly specific, time-dependent predictions regarding how we will look and why we will look that way, are entirely unfounded. Though as is often the case, human imagination trumps restraint, and the desire to speculate commonly wins out over a more cautious approach to investigating these questions.

For instance, in 2013, Dr. Alan Kwan, a computer scientist at Washington University, worked with artist Nickolay Lamm to predict what humans may look like in the distant future. They offered up images of modified human heads, with larger eyes and bigger foreheads, and compiled a list of mechanisms that might lead to the development of these and other specific future human attributes. Additionally, they devised timelines for when and how these may arise during human evolution over the next 20,000, 60,000, and 100,000 years. This begins as a rather benign prediction stating that the human head will trend larger to allow for increased brain size, which is simply a projection of past hominin trends into the near future, without any mention of what may be shaping them specifically. However, once the mechanisms assumed to be responsible for driving these changes are introduced, it quickly turns to a far more speculative pursuit:

> I would hazard a guess that millennia of human space colonization of Earth-orbit and other solar system space colonies will also select for...
>
> 1. Larger eyes in response to the dimmer environment of colonies further from the Sun than Earth ... Eyes would seem unnervingly large to us and have "eye shine" from the tapetum lucidum. Sideways blink of the reintroduced plica semilunaris to further protect from cosmic ray effects would be particularly startling.
>
> 2. More pigmented skin to alleviate the damaging impact of much more harmful UV radiation outside of the Earth's protective ozone.
>
> 3. Thicker eyelids (a more pronounced superciliary arch) to alleviate the effects of low or no gravity that disrupt and disorient the eyesight of today's astronauts on the ISS.[421]

While it is fun to speculate about what types of environmental conditions may contribute to our future evolutionary form—be they earthly or celestial—it is simply impossible for anyone from our time to know for sure. Even if we are able to see the end result of these evolutionary forces in the anatomy of our extratempestrial descendants, we are still unable to know what specific causal factors went into creating those traits between now and then. It is for this reason that I have attempted to steer clear of these types of predictions, and instead, focus on well-established long-term trends that have persisted throughout the hominin past here on Earth.

With specific regard to Kwan and Lamm's predictions, they make the unfounded assumption that it will somehow be possible to travel to, or live in, other solar systems. Though beyond the daunting challenges of space travel and colonization, it is important to recognize that we share a deep phylogenetic relationship with other primates and mammals. As such, new anatomical features are not likely to break longstanding patterns of cladistic homogeny and suddenly arise in only our species. For example, humans—as well as most other primates—no longer have a tapetum lucidum, which is the thin layer of tissue behind the retina that causes the eyes of many mammals to glow when light strikes them. Because the majority of primates lack a tapetum lucidum, we must have lost this trait a very long time ago, and we are not likely to get it back anytime soon.

There is also no sound logic behind why a plica semilunaris would suddenly reemerge as a third eyelid either, given that this trait was lost even farther back in our evolutionary past, as indicated by how rare it is among mammalian species alive today. Rather, there is good reason to believe that neither this, nor any other new trait, would develop in the human future, largely because of our long-standing tendency to solve problems culturally, and not by constantly evolving new anatomical features to cope with new environmental conditions.

This cultural mitigation mechanism has substantially reduced selective pressures that used to be placed upon us, particularly in more recent times, as our culture became increasingly sophisticated and better able to solve more complex problems. So, rather than using genomic engineering, or waiting for natural selection to slowly shift our gene frequency in the direction of greater plica semilunaris or tapetum lucidum development to combat the effects of solar radiation in space, we could simply put on a pair of sunglasses, or perhaps wear some sort of tinted contact lens to protect our eyes.

In all fairness, when attacked for not making a clear enough distinction between speculation and science, Dr. Kwan responded that he was misquoted. Kwan stated that his conjectures were framed by the *Forbes Magazine* interviewer in a way that made them sound as if he was making scientific predictions, when in fact they were always meant to be light-hearted reflections on possible future human scenarios.[422] Nevertheless, the contentious row that arose between Dr. Kwan and the *Forbes Magazine* writer—who fervidly criticized him and his ideas—further highlights the importance of drawing a clear distinction between the known and the unknown and the plausible from the impossible. This is particularly true when dealing with uncertainty associated with the human evolutionary future, given that we are all current prisoners locked within the same narrow cellblock of block time.

It is likely that humans will still exist 20,000, 60,000, and 100,000 years in the future. Throughout this time, we will undoubtedly continue to carry the lion's share of morphological traits that have helped define us as a distinct taxonomic clade, over a much longer period of our past existence. Though as our culture and technology continue to evolve at a faster rate than our biology, it is impossible to say how human anatomy and cognition may be affected, as the two will undoubtedly continue to interact synergistically, just as they have throughout the majority of our evolutionary history.

Indeed, there is no equation in Darwin's *Origin of Species*, or elsewhere, which can precisely predict what a species will look like in its evolutionary future. It is important to recognize this present limitation to avoid unabated speculation about how and why we may eventually come to look or act a certain way. However, it is also important to recognize that past trends can help inform some aspects of our future state. This is particularly true of changes that have transcended innumerable environmental and cultural circumstances and that have endured among multiple different hominin species, spread across the entire length of our branch upon the tree of life.

Humans exist today as an exceptional earthly species and a doyen of the natural world. We have come to enjoy an intellectual sophistication unknown anywhere else on earth or in the universe beyond. In examining the accelerated rate of human cognitive and cultural change, it is certainly conceivable that someday we could obtain a new level of cross-temporal complexity in which elements of human culture, technology, and biology become increasingly integrated across different periods of the past and future. The implications of broad-based intertemporal interaction would certainly be vast. For if there has been, and will continue to be contact among temporally distinct members of the hominin lineage, then we are clearly fated to become an even more advanced species than we have ever been at any point within an already deep and multifarious biocultural evolutionary past.

11

Implications of Intertemporal Interaction I

Early humans possessed a survival trait that was invisible because it was locked within their skulls and only revealed through their behavior. Their immense and complex brains conferred tremendous intelligence, along with a vast capacity for memory, an insatiable curiosity and astonishing creativity—abilities that more than compensated for their physical and sensory deficiencies.[423]

– David Suzuki

Where we are going as a species is a big question. Human evolution certainly hasn't stopped. Every time individuals produce a new zygote, there's a reshuffling and recombination of genes. And we don't know where all of that is going to take us.[424]

– Donald Johanson

11.1. Biological and Cultural Anthropology through Deep Time

Between six and eight million years ago, our ancestors began walking upright. Although this early behavioral shift may seem rather insignificant, it actually had far-reaching implications for the ensuing future of our lineage, as it helped shape a multitude of physical and behavioral traits that went into making us the most advanced organism that has ever existed on this planet. We can learn a lot from investigating the fossil and skeletal morphology of our ancestors, as well as the material remains they left behind. Though we remain limited in the amount we can learn from our restricted vantage point in time, given that the available material dwindles as we venture deeper into the past. However, if we were someday able to travel backward through time, it would afford us an incredible opportunity to learn much more about our rich prehistoric past.

We would no longer be left to scrape through layers of dirt to discover our origins and evolutionary history, as we could instead conduct cross-temporal ethnographies and biomedical examinations of our hominin ancestors, gaining a much clearer understanding of their biology, morphology, and changing patterns of culture and cognition. It is this aspect of abduction reports that are most intriguing, as the processes described are precisely what we anthropologists would do in order to learn as much about the human past as possible, if we too possessed the technology to engage in intertemporal field research. Everything about these visits, as described by those who claim to have been a part of them, seem to indicate that the principal aim is to collect data about the ancestral human past, using technology that anthropologists of the modern era could only dream of.

The anthropological subfield of biological/physical anthropology is discussed throughout this text, including our focus on investigating bipedal locomotion, long-term craniofacial trends, ontogeny, neoteny, and the complex relationship between culture and biology through time. Though another important area of research in biological anthropology involves mapping genetic variation and the frequency of various alleles, meaning alternative forms of a gene, across modern and past hominin groups. This provides information about population origins, migrations, gene flow, gene drift, isolating mechanisms, group replacements, as well as the impact of disease and other forms of natural selection on human populations through time.

By sampling the genotype of individuals across various modern human populations and looking at the global distribution of certain alleles, we are able to test theories about changes to human genetic variability across time and space. This is a very useful tool for researching the past, but it also tends to leave a number of unanswered questions. This is largely due to limitations in our knowledge regarding the specific social and environmental factors that helped shape certain gene distributions. We are also limited in our ability to extrapolate modern genetic data into the past, as we are currently provided only a snapshot of what was actually a long and complex interaction among individuals on a local, regional, and, more recently, global scale.

With further technological advances and the discovery of new hominin specimens, we have been finding and examining samples of preserved

DNA in the bones and teeth of our ancient human ancestors. This helps reveal certain more intricate aspects of our evolutionary history, as well as previously unknown forms of interaction among different groups through time. For example, the discovery of ancient DNA in the remains of Neanderthals and Denisovans—another archaic *Homo sapiens* group with whom we also share a recent common ancestor—has shed light on a number of mysteries of our genetic past. Most remarkable was the revelation that modern humans, Neanderthals, and Denisovans had all been interbreeding with one another, despite the longstanding notion that these three groups were each an entirely different species.[425][426][427] In fact, as a result of the discovery of preserved ancient DNA, we now know that all modern human groups outside Africa retain some amount of Neanderthal DNA, with Eurasian populations carrying between 1% and 4% of it on average.[425]

Some gene sequences from Neanderthals are even thought to confer certain benefits, such as enriching genes that affect keratin filaments, which may have helped our modern human ancestors adapt to variable environments outside Africa.[428] However, certain other regions of our genome are completely devoid of any Neanderthal DNA, which suggests that it may have conferred a risk for disease, or reduced our selective fitness in other ways. For instance, the lack of any Neanderthal genes specifically expressed in the testes and their scarcity across the X chromosome of modern humans indicates that Neanderthal genes may have reduced male fecundity once they were acquired in our *Homo sapiens* ancestors, since the X chromosome contains a higher proportion of genes associated with male hybrid sterility.[428] If true, this would constitute a sort of serendipitous genetic defense against a takeover of the Neanderthal gene pool once modern humans began to colonize the land they had dominated for nearly 400,000 years prior.

Ancient DNA research shows that modern humans were also interbreeding with the Denisovans, even after this Neanderthal admixture, since as much as 5% of Denisovan DNA is still present in some modern human groups, even though *Homo sapiens* males who acquired their DNA may have faced reduced fertility from it.[429] Modern South Asians, Australians, and particularly oceanic populations, have a disproportionately high frequency of these Denisovan genes, which points to more recent admixture between the Denisovans and these modern human populations.[430] However, genetic isolation and a lack of gene replacement by other groups who had a larger share of their ancestral DNA whittled away by gene flow through time, also could have contributed to the higher frequency of Denisovan genes in these populations. So, in the same way we still have egg-laying mammals and marsupials on some of these islands, Denisovan genes could have lasted simply due to geographic and genetic isolation, following earlier instances of interbreeding.

Multiple other factors associated with population dynamics, gene flow, gene drift, mutation, and natural selection in response to localized environmental conditions, also influence the way our human genome waxes or wanes through time. Without a representative sample of the genes of different geographic and temporal groups from equally spaced periods of the past, it is difficult to know what specific mechanisms were involved in shaping patterns of genetic change across time and space. However, if anthropologists were able to travel back to predetermined points in our evolutionary past to collect genetic data from earlier human populations, we would be provided a much richer understanding of the genetic history of hominins on this planet.

As time travelers of the present, we could pick people up and take tissue and fluid samples to run any number of population genetics, evolutionary biology, health, or reproductive studies. We could take fecal samples to understand what types of foods they ate and to examine how our diets have changed over time. We could look at their clothing, tool technology, and various forms of communication to understand the cultural process. Furthermore, if the group had developed a preservable written language that could be studied prior to visiting, it would even be possible to ask questions that would allow for deeper insight into their symbolic lives, including their views on creation, the afterlife, how they perceive their own existence, and how they perceive us as visitors to their time.

Modern humans who claim to have been abducted by extratempestrials most commonly describe a series of events eerily similar to those listed above. Based on the available evidence highlighted throughout this text, such encounters point to instances of intertemporal inquiry, carried out by scientists of the distant human future, tasked with examining obscure facets of their own biological and cultural past. This is also indicated by numerous abduction accounts in which people report being stripped naked and enduring something very much like a modern medical examination. More specifically, during these procedures, blood, skin, and other body tissues are sampled; the eyes, nose, ears, and mouth are examined; and fecal samples are commonly taken, which is what likely gave rise to the infamous "anal probe" aspect of IFO abduction reports. For academics involved in most forms of biological research on human and non-human animals, it is easy to see how these procedures and the data generated from them would provide a wealth of information regarding the human biological condition throughout our evolutionary past.

11.2. Intertemporal Fecundity

In addition to sampling fecal matter, blood, skin, and other bodily tissues, a significant percentage of abductees also report having sperm, eggs, and occasionally even developing embryos extracted from their person. For instance, in a well-documented and well-publicized case, Clayton and Donna Lee—a husband and wife from Houston, Texas—report having been abducted several different times. On one occasion in particular, Donna recalls having a gestating fetus removed from her uterus by individuals she described as "tall, slender, blonde," and who "looked human but not human, not quite human; they didn't really use their voices to talk."[431] A 2006 video interview conducted by Bill O'Reilly on Fox News lends some credence to this testimony. The Lee's body language, tone, and spoken dialog suggests that neither of them ever wanted this, nor did they want to be talking to Bill O'Reilly or anyone else about it, had they not suffered through a nightmarish experience that made them desperate for answers.

Other women have also described situations in which they were abducted, impregnated, placed back in bed with a clouded memory, and then abducted again later, at which point the developing fetus was extracted from their bodies. Still others have reported being given a small child to hold while aboard an IFO and told that it was a baby they had helped to create. It should be noted that in addition to the horrors of being abducted, undressed, examined, and enduring unsolicited touching, tissue sampling, and gamete extraction, being given a baby and being told that it is yours would also certainly cause intense stress and anxiety among anyone subjected to such unethical treatment.

Men commonly report having to endure naked and humiliating fertility procedures as well. For instance, male abductees often describe a jockstrap-like cup being placed over their genitals, which causes them to ejaculate almost instantly. Such a device, devised to elicit and capture semen samples, is not inconceivable either, and particularly considering that something very similar to this has already been developed and introduced to a hospital in Nanjing, China.[432] This novel contraption allows men to produce a "hands-free" sperm sample, and though it is ostensibly more primitive compared to the instant-ejaculator of our extratempestrial descendants, it does perform the same function. Furthermore, the modern existence of this same technology, developed for the same purpose, makes reports of its future reality seem more credible, as its existence now and in the reported future denotes cross-cultural continuity through time.

In addition to cases of gamete extraction, some abductees have even reported participating in actual intercourse with their fourth dimensional abductors. One of the best known and most credible accounts comes from the 1957 abduction of a reputable Brazilian lawyer named Antonio Villas-Boas. To avoid the intense heat of the day, late at night on October 16, 1957, Villas-Boas, a 23-year-old farmer at the time, was ploughing a field when he noticed a red light in the sky. He states that it moved toward him, at which point it became recognizable as a large egg-shaped craft, with red lights toward the center and a rotating dome-like structure at the top.

After being abducted by humanoid beings aboard this craft, he reported being stripped naked, washed, and then laid down on a table. Shortly thereafter, a female extratempestrial—with blond hair, blue slanted eyes, a wide face, thin lips, and a very pointy chin—came into the room and had sex with him. This account, while certainly strange, would also be expected to generate a high reliability rating on Hynek's scale, given that this successful attorney, of sound mind and body, vehemently stood behind this claim until his death in 1991.[433][434]

The importance of gamete procurement and zygote production in instances of intertemporal interaction is also signified by the fact that nearly all of those abducted are healthy, under the age of 40, and are reproductively viable. The latter of which is indicated by a specific case in which a man is reported to have been rejected by "the aliens" because he had previously undergone a vasectomy.[435] Cross-temporal reproductive tissue sampling, in vitro fertilization, and interbreeding, if true, could have immense implications for the genetic future of our evolutionary lineage. Additionally, based on past and current trends in human population genetics, intentional management of the future human germ line may even become a necessity of our distant descendants.

11.3. On the Banks of the Deep End of the Hominin Gene Pool

A number of long-term morphological and cultural trends that link the hominin past to the present condition of our species are examined throughout this text. A principal aim of this inquiry is to demonstrate how the continuation of these enduring trends could eventually result in us becoming the "aliens," which have thus far been erroneously regarded as creatures from another planet. These long-term biocultural trends in the hominin past, extrapolated into the future, suggest that these are indeed aliens of time, rather than space. The development of a larger and more bulbous brain, reduced face, reduced body hair, and other changes in hominin evolution, establish continuity through time and a strong phylogenetic link among humans of the past, present, and future. However, no other characteristic—physical, cultural, or otherwise—could provide more direct evidence of shared ancestry than our capacity to reproduce with each other, as this is and has long been the gold standard used to determine if two organisms are in fact the same biological species.

Simply stated, the Biological Species Concept (BSC) asserts that if two separate organisms can mate, reproduce, and produce viable offspring—meaning that their offspring is also able to reproduce itself—then they are considered to be of the same species. The most common example of this involves the mating of a donkey and a horse, where the result is a living breathing hybrid known as a mule. However, mules themselves are always infertile, thus making the horse and donkey two separate species using the BSC criteria. If there is truth to the plethora of reported instances of gamete extraction, zygote formation, interbreeding, and most importantly, the production of viable offspring among humans from vastly different time periods, this would not only point to a phylogenetic link among us, but it would also demonstrate a shared species designation among hominins separated by vast swathes of time.

11.3.1. The Question of Species Designation in the Past and Future Hominin Record

A principal quandary in paleobiology has to do with determining the best way to classify the vast array of organisms that have evolved from and into each other over the 3.7 billion-year history of life on this planet. Macro-evolutionary changes occur over immense time periods, and in the process of species giving rise to new species, a great number of changes to the physical form of an organism can occur. Conversely, some organisms have maintained a relatively constant phenotype over an extended period of time, such as the 390-million-year-old coelacanth or the 450-million-year-old horseshoe crab. This variable rate of change, which fluctuates in association with numerous internal and external factors, adds to the problem of species designation in the fossil record. In fact, defining a new species in relation to similar other ones generally equates to drawing an arbitrary line across a long and unbroken gametic continuum.

New species arise through the bifurcation of an existing interbreeding population, whether it be through a sheer divergent split or one group carrying on as another buds off from it incrementally over time. In either case, the separation of some members of a gene pool from other members of the same gene pool can lead to reproductive isolation. Over time, sustained impediments to interbreeding, accompanied by divergent evolutionary forces acting upon what was previously a singular group, causes them to bifurcate further. Over even more time, these forces act to create new and different species, to the extent that neither group recognizes the other as a viable mate. Or, depending on how much change has occurred since the split, these descendant groups may not be able to mate and reproduce with each other at all, because their physical form or genetic structure no longer allows it.

Paleontologists struggle with accurately designating past species as a result of difficulties in knowing how long it takes for speciation events like this to occur, especially because it has shown to be highly variable among plants and animals across time and space. Other factors also complicate this process, including certain sparse parts of the fossil record, difficulties in assessing how much geographic, sex, age, and other forms of morphological variation exist within and among temporal groups, and how best to account for each of these in order to definitively say that one specimen is indeed a separate species from any others.

We must also consider the multitude of factors that contribute to phenotypic variation in assessing whether newly discovered hominin fossils should or should not be classified as a different hominin species, particularly considering that morphological variation is not always representative of genetic variation. For instance, based on their skeletal anatomy, a wolf and a Chihuahua would undoubtedly be considered two different species by future paleontologists that happen to dig them up. In fact, they would likely even be considered a separate genus under the Morphological Species Concept (MSC), which, as the term implies, groups species based on morphological affinities when it is unfeasible to examine their genetic makeup or to assess their ability to reproduce with one other.

However, because a wolf and Chihuahua are physically able to produce viable offspring, albeit rather awkwardly one would assume, under the Biological Species Concept they would in fact be considered the same species. So, in spite of thousands of years of behavioral and genetic isolation—not to mention highly divergent natural and artificial selective forces being applied to their respective genomes—the mighty wolf and the meager Chihuahua both retain the *Canis lupus* epithet. In a recent interview for *Livescience*, Jack Tseng, a paleontologist at the American Museum of Natural History, was asked: *If aliens visited Earth tomorrow, would they realize that dogs—from the spotted Dalmatian, to the giant Great Dane, to the tiny Chihuahua—are all the same species?* to which he responded:

> Forget aliens, if we hadn't actually bred dogs ourselves, even humans would have a hard time determining that a Cavalier King Charles spaniel and a wolfhound are related....If you were a biologist who comes from a society that never had any dogs associated with humans and you looked at these dogs, you would immediately think that these were different species.[436]

Modern humans also possess a number of distinct traits that differentiate us from others throughout the world. These observable physical variants are the result of adapting to different environmental conditions over recent evolutionary time. Certain phenotypic differences among modern human geographic groups can appear to be somewhat stark, though the amount of genetic variation among living people is actually quite small relative to most other animals.[437][438] Additionally, regardless of how dissimilar people from different parts of the world may look, we still maintain the ability to reproduce viable offspring with one another, as we have for thousands, and possibly even hundreds of thousands or millions of years.

The relatively high degree of observable physical dissimilarity among modern dogs, and to a lesser extent modern humans, must also be considered when assessing how accurately we can designate hominin species in the fossil record. In fact, if we look at variation among fossil hominins representing different geographic regions over the last 1.5 million years, there is still far less variation among us, compared to the highly divergent, yet still reproductively viable members of *Canis lupus*. It is only recently that we have begun to include the Neanderthals as a subspecies of humans (*Homo sapiens neanderthalensis*), and even then, only after genetic evidence clearly demonstrated that we were procreating with them (BSC), despite being quite different morphologically (MSC).

Comparing the cranial morphology of *Homo erectus* to that of modern-day *Homo sapiens*, it is clear that a number of differences exist, most notably with regard to neurocranial and facial size, skull thickness, and overall craniofacial shape. If the multitude of images and descriptions of our extratempestrial descendants can be used to inform future hominin morphological differences, it could be argued that a similar degree of morphological variability exists between us now and us in the future. In fact, depending on the specific time period from which they come, one could argue that the craniofacial differences between extratempestrials and ourselves are somewhat fewer than those of modern humans and *Homo erectus* 1.5 million years ago.

At present, we are unable to test whether we could produce viable off-spring with *Homo erectus* 1.5 million years in the past. Though indications of an ability to reproduce with our distant descendants—who are characterized by a similar degree of morphological divergence—may point to a long period of reproductive viability throughout our distant past as well. However, until we become able to traverse time to test the Biological Species Concept ourselves, species designation in paleoanthropology is likely to remain a somewhat contentious affair.

If gamete extraction and concurrent coitus are indeed common components of extratempestrial contact, it not only points to our shared species status, but also raises the equally important question, why? It is certainly in our nature to want to have sex and we have a long history of getting it on with those both like and unlike ourselves, such as during missionization, colonization, military conquest, and in instances of interbreeding between *Homo sapiens* and Neanderthals as discussed earlier. Although, because we are slowly becoming one large interbreeding population on the isolated island of Earth, other more functional, population genetics-based facets of this odd form of intertemporal interaction may also need to be considered.

11.3.2. Genetic Homogenization and Intertemporal Interbreeding

Throughout the majority of hominin history, our ancestors inhabited only a tiny sliver of the vast expanse of dry land on this planet. However, beginning around 2 million years ago, when *Homo erectus* first migrated out of Africa, we began to occupy a wider portion of the Old World. Over time, as we continued to saturate these previously unpopulated parcels, voyaging into what would become the Americas around 20,000 to 30,000 years ago, we came to inhabit nearly every landmass on planet Earth. As we filled these new territories with ever-increasing population sizes and densities—especially following the Neolithic revolution and the dawn of agriculture—we began to migrate and mate over much larger territories. Though even as our populations grew, natural and cultural boundaries restricted gene flow among them, and the enduring forces of gene drift, mutation, and natural selection helped maintain a relatively high degree of genetic variability among more widely dispersed geographic groups.

However, a trend toward largescale human genetic homogenization was set in motion beginning around 1,000 years ago with Viking explorers, and more overtly around 500 years ago, as most of the rest of Western Europe began colonizing more diffuse regions of the world. Today, with the ability to board an airplane and make a baby with anyone, anywhere in the world in less than 24 hours, we continue to chip away at longstanding localized gene variants and move toward becoming an amalgamated global human population. In essence, advancements in transportation technology have slowly drained these formerly dispersed gene pools into one large genetic ocean, as we've trended toward becoming a singular interbreeding intercontinental population.

Large-scale homogenization of the entire human genome could present some problems, if this trend were to continue into our evolutionary future…as it is likely to do. Most evident is the degradation of genetic variation, which is vital to maintaining the adaptability and survivability of any organism, as per the laws of evolution. For example, decreased genetic variation in the human leukocyte antigen system (HLA) that encodes the major histocompatibility complex (MHC), which together form the backbone of our immune system, would make us more susceptible to diseases, which could potentially wipe out large segments of the human population. These critical components of our immune system have evolved to become some of the most variable parts of the human genome, as this genetic variability helps protect us from an array of dangerous pathogens.

In fact, this overlooked aspect of the evolutionary process was one of many logical, logistical, and ethical flaws behind the eugenics movement, and particularly with regard to the Nazi philosophy of creating a "super-race" of genetically "pure" humans. This is for the simple fact that natural selection works on existing variation within a population, and if it becomes too limited, the population as a whole is at risk of complete annihilation. Instead, a broad and diverse population, characterized by a high level of genetic variability, is beneficial for the entire group.

However, if past and current trends persist, and we continue to move toward large-scale homogenization of the human genome, we may ultimately be forced to look elsewhere—or more aptly elsewhen—for novel gene variants. Because we cannot simply go to a different planet to sample gametes from other humanoid lifeforms—as extraterrestrial creatures would undoubtedly be too different, and too far away…if they could be located at all—taking a dip into the gene pool of the past may be the only viable option for bolstering future human genetic variation in an effort to stave off a state of global incest.

In the most basic sense, a population is defined as a group of inter-breeding organisms, which are all the same species and which live in the same place and at the same time. If we ever reach a point in our genetic future when we can no longer find concurrent conspecific mates who don't share alleles identical by descent, introducing genes from members of our same species but from a different time may become the only viable option for increasing genetic diversity. From this perspective, the accelerating rate of intercontinental interbreeding over the last 500 years, among other factors, could help explain why reports of gamete extraction are so common among those who claim to have been abducted by our distant descendants. In other words, if this trend toward global genetic homogenization were to continue into the deep future, stealing sex cells from asynchronous and more genetically heterogeneous groups of the past may be critical to refining an otherwise stagnant future human gene pool.

It should be noted that mutations have consistently added new genetic variants to life on this planet. Mutations help increase variation within an interbreeding population and are an important contributor to macroevo-lutionary change through time. Though while new mutations may help increase genetic variability, most are bad, and detrimental alleles would not be expected to circulate widely throughout a population. Furthermore, even if a new mutation arose that coded for something advantageous, it may eventually spread throughout the entire population and therefore not do much to reduce overall genomic homogeneity.

Interestingly, research has shown that it does not take a lot of new gene variants to cancel out the negative effects of inbreeding. So, rather than setting up 24-hour sex stations spread across different periods of the future human past, modest sampling efforts—such as those that charac-terize abduction reports—could be enough to mitigate the undesirable consequences of broad-based intra-planetary incest. As a matter of fact, this is something that modern humans already do to other animals, as it is common practice in plant and animal conservation projects to in-troduce novel gene variants from other populations, which helps bolster genetic diversity among groups that have suffered size reductions and subsequent genetic homogenization.[439] These efforts help restore genetic heterozygosity and reduce the negative effects of small population sizes, gene drift, and inbreeding.

For some time now, largely thanks to the work of legendary geneticist Sewall G. Wright, we have been able to quantify the level of heterozygote deficiency within a population. Wright's F-statistic, or *fixation index*, provides a statistically expected degree of reduced heterozygosity in subdivided populations relative to one in Hardy-Weinberg equilibrium, meaning a population in which allele frequencies are constant, and therefore, no evolution is taking place.[440][441][442] Wright also introduced three interrelated variables to measure the amount of heterozygosity at various levels of the subdivided population structure…F_{IT} is the correlation between gametes in an individual compared to the entire population; F_{IS}, also known as the inbreeding coefficient, describes the correlation between gametes within an individual in relation to their subpopulation; and perhaps the most valuable parameter among them, F_{ST}, is the correlation between gametes within subpopulations relative to gametes drawn from the entire population.[443]

Values for F_{ST} range from 0 to 1, where a low F_{ST} designates a higher level of shared genetic material and a lot of breeding, while a high F_{ST} means fewer shared genes, which is often the result of low migration rates and relative geographic or behavioral isolation. In addition to these mechanisms affecting population genetic structure, Wright also provided a thorough understanding of how mutation, gene flow, gene drift and natural selection influence genetic differentiation among populations.[444] In this capacity, Wright's F_{ST} has proven to be quite useful, particularly in studies of human genetic diversity.

As stated earlier, genetic diversity among modern human populations is quite low,[437][438][445] which is related to a number of evolutionary and demographic factors in the human past. Although it is currently impossible to know the value of F_{ST} at various stages of our evolutionary future, taking into account our already low fixation index, as well as numerous biocultural trends in the recent past—and especially rampant intercontinental gene flow over the last 500 years—it is very likely that F_{ST} will continue to decrease through time. Considering this trend toward global genetic homogenization, it is not unreasonable to assume that abduction reports describing the extraction of modern human semen, eggs, developing fetuses, and even direct sexual intercourse, stem from efforts by our extratempestrial descendants to manage, or perhaps mitigate problems associated with reduced genetic diversity in the distant human future.

Coincidentally, while doing interviews with various media outlets to promote this book over the last few months, I was told on multiple occasions about the experiences of an individual named Sergeant Jim Penniston. Penniston was a serviceman at the RAF Woodbridge Air Force base located in Suffolk, England, when an alleged IFO encounter took place in the Rendlesham Forest near the base. Late at night, on December 26, 1980, while investigating these strange lights, Penniston claims to have touched an IFO, at which point he received information about the craft, its occupants, as well as details regarding their purpose for being here. Later, under hypnosis, Penniston recalled being told by these extratempestrials that they are time travelers from our future and that they are returning to the past to obtain genes and chromosomes to help maintain the future human population, which is having problems with reproduction.[446]

This encounter is notable. Not only because it corroborates the current time-travel hypothesis—not to mention that it is from a fundamentally different source with very different reasons for arriving at this conclusion—but also because it highlights the research aspects of these visits, as well as the role that gamete extraction may play in helping to mitigate problems of limited genetic diversity among our distant human descendants. Obtaining DNA from members of the hominin past makes sense in the context of our ongoing trend toward becoming one global human population, particularly if the sampled gene variants were not ones that persisted into the future to become a part of the broad-based extratempestrial genome.

In addition to global migration and an accelerated rate of gene flow among recently distinct human subpopulations, other factors, such as a population crash resulting from global war, environmental change, natural or manmade disasters, asteroid impacts, climate change etc., could further reduce heterozygosity in the human future. In fact, there are precedents for these kinds of genetic bottlenecks in human history—a part of why we currently possess a relatively low F_{ST} value—which includes a major decline in human genetic diversity that occurred as recently as the Neolithic.

Beginning around 7,000 years ago, and for about 2,000 years after, human genetic diversity plummeted throughout Europe, Asia, and Africa. More specifically, Y-chromosome diversity decreased substantially, which indicates that the bottleneck was mostly tied to men. This Neolithic

Y-chromosome bottleneck was somewhat of a mystery to anthropologists, until recently at least, when a precocious undergraduate sociology major named Tian Chen Zeng offered an insightful explanation for it.

Zeng recognized that the Neolithic was characterized by widespread patrilineal descent, in which land, wealth, family names, etc., were passed down through the paternal side. There was also a lot of war and conflict over resources following the advent and spread of agriculture, which resulted in a disproportionate number of men being killed off in conflicts among rival patrilineal clans. This left relatively few men in relation to women and relatively few powerful and prolific patrilineal clans composed of men who were all related and who therefore shared the same Y-chromosome. Over the next 2,000 years, as these genetically homogenous patrilineal clans grew and expanded, it put limits on overall human genetic diversity.[447]

As humans move toward becoming one large interbreeding population, if there were to be an event that further limited our genetic diversity beyond that which followed the Neolithic revolution, it could put our distant descendants in a bit of a genetic bind. Additionally, cultural processes, fertility trends, and even intentional attempts at manipulating human population sizes, could potentially result in additional unintended genetic consequences. For instance, the recent global trend toward reduced fertility, largely associated with changing demographics and more opportunities for women in the workplace, has resulted in many nations falling below the replacement rate of two children per couple. The one-child policy in China represents another more institutionalized attempt at limiting fertility, which had an unintended effect on the sex ratio in that country, to the extent that China has long-held the most skewed sex ratio at birth of any nation.[448]

An acceleration in the speed and scale of global travel over the last 500 years has also acted to blend the overall human genome, and at a faster rate no less, while a recent increase in international adoptions further blends formerly distinct human subpopulations. Considering these and other biological and cultural processes in recent human history, additional limits on human genetic diversity in the distant future could potentially force our descendants to go in search of new gene variants, in the only place a more diverse gene pool ever existed, the hominin past.

There are certainly a multitude of other potential explanations for why these reports of gamete extraction are so common in IFO reports. Though based on past and ongoing trends in human and non-human population genetics, as well as the ardent testimony of Sergeant Jim Penniston, it could very well be associated with forthcoming efforts to lessen the detrimental effects of a seemingly inevitable state of global inbreeding. A future problem that may be circumvented by flowing novel gene variants from more heterogenic human populations of the past into more homogenic human populations that lie to our distant future.

11.4. Interfacing Time-Races—Physiology, Culture, and Communication

In spite of past and current trends toward global human genetic homogenization, future humans are certain to retain some level of morphologic, cultural, and linguistic variation. As such, we must consider the implications of time travel on our conceptualization of human variation, both within and among different time periods. That is, with the advent of a device capable of achieving backward time travel, an individual's observable cultural and physical traits would need to be assessed in the context of their temporal origins, just as we currently consider cultural and biological variation among modern humans in the context of their geographic origins.

As is clear to anyone who travels internationally, people look different in different parts of the world, which is more evident the farther we voyage out away from home. The same could be said for anyone traveling outward away from their home time, in either direction. This is because phenotypic differences become increasingly divergent among human groups representing increasingly disparate time periods, due to evolutionary forces having longer to act on these temporally distinct populations.

As noted above, many modern humans who experienced a close encounter with future humans described the variation among these extratempestrials as being similar to what we currently observe among ancestral groups, or what was previously referred to as "race." This commonly reported phenomena could be interpreted as the persistence of geographic races into the future—if all those aboard were from the same period—or these extratempestrial disparities could reflect intertemporal variation, which may suggest collaboration among researchers from different periods. In the latter context, ambassadors of different time periods would possess phenotypic characteristics representative of the specific era from which they came, or what could be considered *temporal ancestry*.

This aspect of time travel would add further variation to that which presently exists among geographic human groups and would act to broaden the observable range of physical variation among any group that lies within the temporal reach of a time machine. For example, consider visiting a patisserie in the Latin Quarter of Paris each morning for a baguette and some éclairs; over time, the owner may assume that you live in the area and enjoy delicious baked goods. If you had a foreign-sounding accent, or some other non-French characteristics, clothes, customs, or behaviors, the shop owner may assume that you were originally from somewhere else, as this is a conventional concept in the context of constant contact with concurrent conspecifics.

No matter how different you may seem to them, this Parisian patisserie owner is not likely to consider the possibility that you actually hail from a different time, given that this is not yet a common aspect of our shared reality. However, if they started seeing people come in with varying degrees of neurocranial globularity, protruding foreheads, retracted faces, huge eyes, further variation in skin tone, chin shape, amount of hair, and new accents that don't fit with any existing linguistic group, they may be forced to consider that their patisserie has recently become an intertemporal phenomenon.

11.4.1. Intertemporal Linguistics

During the early 20th century, cultural anthropologists began to realize that the best insight into a culture is achieved by learning the language of the specific group being investigated. This important focus on language and communication spurred a revolution in the anthropological sciences, which largely began with Franz Boas in the early 1900s. Regarded as the father of American anthropology, Boas emphasized learning the language of a cultural group prior to beginning ethnographic fieldwork with them. This emphasis stemmed from the fact that our attainable knowledge of any society increases as we gain the ability to communicate directly with them. Language is an inseparable element of every culture, and in order to truly understand the idiosyncrasies of a society, one must first understand their language.

As stated previously, the field of anthropology is divided into sub-fields, or specializations. This is largely because humans are so complex that we require a broad approach involving multiple specialists working together to create a holistic understanding of our species across space and through time. It is common for cultural anthropologists, archaeologists, and linguists to specialize in specific geographic regions or time periods, which facilitates deeper understanding of human lifeways at those places and times. In fact, many linguistic experts are capable of translating languages that have been out of use for hundreds and even thousands of years.

An appreciation of our modern ability to learn languages of the ancient past helps frame our distant descendant's reported ability to speak our temporally indigenous languages of their own ancient past. In fact, our modern languages would likely be much easier to learn by comparison, considering that we leave behind more cultural materials that could aid in their efforts to understand them. In fact, we go to great lengths to archive and protect important cultural documents. If these attempts to preserve the past prove to be effective long into the future, it would be relatively simple for our distant descendants to learn our various modern languages, regardless of how long ago they were spoken by the time we develop time travel technology.

As with most other elements of culture, languages evolve extremely quickly relative to the slow rate of biological change. As such, people from the future who are granted the opportunity to visit their ancestral past, probably do not speak the same language as the ancestors they intend to visit. In fact, even relatively recent versions of the same language can be extremely difficult to understand. This aspect of linguistic evolution was discovered during an attempt at reading the original version of Darwin's *On the Origin of Species* in an Evolutionary Theory class in graduate school. Even though Darwin wrote this book less than 160 years ago, in English, my native tongue, it was utterly incomprehensible. Undoubtedly, our modern languages will also continue to evolve and will surely not be recognizable 10,000+ years in the future, though we are certain to retain the ability to learn them, as the need arises.

Our distant human descendant's perceived ability to learn and communicate in the various languages of their ancient past is impressive. Though what is perhaps most intriguing is *how* they do it. Specifically, in nearly every instance of intertemporal interaction where direct communication is reported to have taken place, it was done telepathically. Just considering the few examples from earlier in this text, abductees had stated that "I could hear in my mind a voice saying 'there is nothing to worry about.' I could hear him talking in my mind;"[449] from the case of Clayton and Donna Lee, Donna is quoted as saying that her abductors "didn't really use their voices to talk;"[431] and in the 2003 Florida abduction case the abductee stated "Once again I heard murmurs in another language, then it sounded like they said it inside of my head, but a soft voice whispered 'We're taking you home.'[450]

Reports of "mind-talking" and telepathic communication are exceedingly common among those who have experienced a close encounter with our highly derived human descendants. In fact, these reports are so prolific that in an actual, albeit somewhat comical survival guide titled *The Complete Worst-Case Scenario Survival Handbook*, authors D. Borgenicht and J. Piven caution that if you encounter an "extraterrestrial biological entity" the first thing you should do is control your thoughts, because they "may have the ability to read your mind."[451] Based on the ubiquity of accounts provided by those who claim to have communicated with members of our future human lineage, it would seem as though our distant progeny may eventually develop the ability to communicate telepathically. From our current standpoint in time, it is obviously difficult to know how this observed capacity for telepathic discourse may ultimately arise. However, we may once again be able to look toward our own biological and technological past for some insight into the possible origins of this odd ability.

11.5. Telepathy as a Future Form of Human Communication?

11.5.1. Current Research on Brain-to-Brain Communication

The field of brain-to-brain interfacing (BBI), which entails the real-time transfer of brain activity between living organisms, has seen a number of advances in recent years. For instance, the feasibility of a computer-mediated brain-to-brain interface was recently demonstrated in a study that linked the central neural functions of a human and a Sprague-Dawley rat. In this rather odd yet revealing experiment, human researchers were able to move the tail of an anesthetized rat…with their minds. This tremendous feat was accomplished by means of a transcranial focused ultrasound (FUS) and a computer mediating device. Together, these instruments allowed researchers to move the rat's tail with $94.0 \pm 3.0\%$ accuracy and with an average time delay of 1.59 ± 1.07 seconds between their original thought initiation and the actual tail movement.[452]

Another recent experiment demonstrated for the first time that BBI and the direct transmission of information can be achieved in humans using non-invasive means.[453] In this study, electroencephalography (EEG) recorded the intentions of one subject—dubbed "the Sender"—while transcranial magnetic stimulation was used for delivering this information via the internet to the motor cortex of a second subject, designated "the Receiver." While the Sender could not dictate exactly what the Receiver was going to do, results demonstrated that the Sender was able to cause a desired motor response in the Receiver, to a greater extent than what could have occurred by chance alone.[453]

While it may seem farfetched to think that we could ever directly communicate with one another using only our minds, a technologically mediated system hatched out of current research in this area could potentially help facilitate this end. There is also no denying that this is a rapidly developing field, which is indicated by a marked increase in the number of technology patents specifically related to reading brainwaves.[454] For instance, between the years 2000 and 2009, fewer than 400 of these types of neuro-technology patents had been filed in the United States; however, this number doubled to 800 in 2010, and it doubled again to 1,600 in 2014.[454]

It should be noted that it is becoming common practice for companies—pejoratively dubbed "trolls"—to generate vaguely worded and overly generalized patents, in the hopes that they can later be used to extort funds from actual researchers who hold actual claims to their intellectual property.[455] Though even considering this recent rapacious capitalistic trend, the rate of increase in neuro-technology patents is a testament to growing interest in this area of research and to the future of brain-to-brain interfacing in general.

We cannot currently know whether the prospective future human ability to communicate telepathically is the result of technology implanted in the extratempestrial brain, or if it represents further development of the brain's already impressive functional capacities. Although, considering the evolution of cerebral and craniofacial morphology, it is very possible that this exceptional faculty for telepathic communication may be a product of the brain itself.

For instance, even though we modern humans share the same biological brain structures as our distant kin, we do not have devices in our brains that enable telepathic communication with those that do. However, during alleged close encounters in which telepathic communication was reported to have taken place, contemporary humans could still "hear" and "speak" using this non-verbal form of direct communication, which suggests that no super-cerebral device may be required. Furthermore, the continuation of the same 6-million-year-old cranial and facial trends in hominin evolution could ultimately result in reduced vocal function and the eventual need for a new non-verbal form of communication among humans of the distant future.

11.5.2. The Brain, Face, and Vocalized Speech

Continued advancements in medical technology, coupled with a better understanding of the inner workings of the brain, may someday result in our gaining the ability to implant a device within the human mind that would allow for direct inter-cerebral communication. However, the alleged ability of our extratempestrial descendants to communicate telepathically with us could instead represent yet another anatomical evolutionary tradeoff associated with encephalization in the hominin lineage. It is possible that the same longstanding trends of brain expansion and facial retraction, which have been linked to the reduced function of other skull components as described previously, may also limit our capacity for vocalized speech in the distant future. In other words, if these morphological trends were to continue as they have for the last six million years, it could further limit available space for speech-related tissues, and potentially contribute to the eventuality of exclusively cognitive communication.

Functional and structural features have competed for space within the confined region of the skull throughout hominin evolution. Thus far, our brains have clearly been the winner in the majority of these bauplan battles. Though if our brains are to continue expanding and/or changing shape in association with our faces becoming even more reduced and retracted, this tradeoff could further limit ontogenetic time and space available for proper development of the pharynx, larynx, hyoid bone, mandible, nasomaxillary complex (facial block), and other hard and soft tissues critical to coherent vocalized speech.

Our exceptional ability to intone a greater range of sounds compared with other animals—including our closest primate relative the chimpanzee—is largely the result of changes to our brain, facial, and vocal tract anatomy, which occurred in association with a shift toward habitual bipedalism beginning around 6–8 million years ago. Bipedalism is associated with changes such as cranial flexing, increased cranial capacity, expansion and anterior movement of the frontal and temporal lobes, and a reduction and downward rotation of the entire facial block. These mid and upper craniofacial changes are also tied to downstream modifications to the lower face. These include increased curvature and elongation of the pharynx, trachea, and esophagus, as well as a more V-shaped dental arcade, compared to the U-shaped configuration of the upper and lower teeth in our hominoid cousins and early hominin ancestors.

Our V-shaped dental arcade arose during more recent stages of hominin evolution in association with a widening of the posterior maxilla and mandible. This broadening of the rearward portion of our mouths helped accommodate structural and functional features of the face, which are important for eating, breathing, and speaking. Apes and early hominins possessed a more U-shaped maxilla and mandible in association with their relatively long faces, and smaller and more posteriorly located brains. However, as we stood upright and grew larger and more forward-projecting brains, while our skulls rotated downward and our faces retracted, space became more limited in the region of our pharynx and oral cavity. Widening of the maxilla and mandible in association with this shift toward a V-shaped dental arcade, helped accommodate these vital structures in this anatomical region of the hominin skull.

Also during more recent stages of human evolution, we began to develop a more caudally located tongue, hyoid bone, and larynx—meaning these features became positioned farther down in the throat. At birth, these structures are situated as high in the throat as in other mammals. However, they begin to migrate downward relatively early during growth and development in humans.[456] The result of this unique pattern of ontogeny is that the adult human supralaryngeal vocal tract forms a double resonator. This, along with our remarkably limber tongues and more advanced brains, grants us the ability to produce a much broader range of sounds and pitches in relation to any other mammal.[457][458]

Once this basic anatomical configuration was in place, a synergistic relationship between the brain and language began to develop. This is to say that our sophisticated form of vocalized speech is both the result of, and a contributor to, the accelerated rate of encephalization and facial orthognathism that characterizes the last one million years of hominin evolution. However, and perhaps somewhat paradoxically, if these same long-standing trends were to continue into the distant future, the same traits that contributed to our advanced form of articulated speech could also act to limit it, particularly if these mitigating mechanisms of the past were to someday reach their physiological limit.

Thus far, the negative effects of many evolutionary tradeoffs in this anatomical region of the hominin skull have been allayed, such as with the aforementioned mediolateral expansion of the posterior maxilla and ascending ramus of the mandible, for example. This broadening of the posterior face helped maintain adequate space for structures important to chewing, swallowing, breathing, and speaking, even while our mid- and lower-faces were becoming smaller and more retracted in association with increased neurocranial globularity in recent human evolution.[459][460][461]

Continued widening of the posterior maxilla and mandible may only be able to allay added cranial expansion and facial reduction to a limited extent. In fact, the persistence of this accelerating six-million-year-old trend could ultimately constrain the nasopharynx, oropharynx, hypopharynx, and larynx to such an extent that we may lose the ability for vocalized speech, though undoubtedly while still retaining the much more primal function of breathing, eating and swallowing. If a loss of vocalized speech were to occur in conjunction with our enhanced and expanded brains evolving the capacity for non-vocal brain-to-brain communication, in effect, nothing would be lost, but much could certainly be gained in the context of enriched cognition and cerebral exchange.

Our advanced linguistic facilities were, and continue to be, a vital contributor to hominin biocultural plasticity and our overall success as a species. As such, it is not at all likely that human communication would ever simply disappear. Rather, the same brain that has long-fostered this unique derived form of interpersonal exchange could simply extend that role in a different capacity. In other words, if we were to someday evolve the ability to directly communicate with the brains of others, we would retain an essential element of our past existence, while still allowing for further cerebral and cognitive enhancement, and perhaps an even greater understanding of language, communication, and the human mind.

As with other evolutionary tradeoffs related to our unconventional cerebral and craniofacial configuration, the voracious brain could continue its no-holds-barred takeover of the human skull. Neighboring features that stood in its way would continue to retreat ahead of this evolutionary anatomical blitzkrieg, as they have throughout much of the hominin past. Additionally, because we are an incredibly intelligent species, we also possess the ability to minimize many of the negative effects of these and other morphological impasses by dint of our highly advanced culture; adverse anatomical traits, which in and of themselves, are a paradoxical byproduct of our colossal cerebra.

The radical transition in size, shape, and processing power of the human brain has afforded us numerous avenues for success throughout the last six-million-years of hominin evolution, despite a cornucopia of anatomical and sociocultural complications that arose in association with it. In light of these challenges and successes, the novel development of a telepathic form of communication, as indicated by numerous alleged instances of inter-temporal interaction, may represent yet another example of how our run-away brain can both create and solve its own morphological quandaries.

12

Implications of Intertemporal Interaction II

All time is all time. It does not change. It does not lend itself to warnings or explanations. It simply is. Take it moment by moment, and you will find that we are all bugs in amber.[462]

– Kurt Vonnegut

12.1. Compounding Cyclic Cultural and Technological Change

Language and communication represent a small but exceedingly important element of our highly evolved cultural knowledge. Since the origin of preservable material culture, beginning around 3.3 million years ago with the advent of simple stone tools, the speed of technological change has increased exponentially. This rapid rate of change is now an accepted part of life in the industrialized world, to the extent that what is depicted as futuristic science fiction in television and film commonly becomes blasé reality in only a few short decades. To put this in perspective, only 66 years separates the Wright brothers' first controlled, powered airplane flight at Kill Devil Hills, North Carolina in 1903, from when we landed on the moon in 1969.

This hasty pace of cultural change can seem overwhelming at times, particularly as we grow older in life. Many find it easier to simply drop out than to try to keep up with the ever-quickening pace of technological change, opting instead for the simpler and more familiar tools of their youth. By contrast, younger members of a society tend to embrace this change and enthusiastically seek out novel products and new ideas. In fact, a comprehensive understanding of new innovations is often a requirement during many stages of life, simply to stay competitive in a rapidly changing job market and as part of educational, political, and social affairs.

It is common to hear complaints about the fast pace of cartoons, video games, and even everyday life, and that these aspects of modern media are overloading our children's fragile little minds, causing irreparable damage to their psyche and ability to focus on one specific task for longer than a few seconds at a time. However, it is this fast pace and rapidly evolving rate of change that allows their incredibly plastic brains to acclimate to the reality of the time period they were born into, while also setting a new standard that helps define the rate of change among subsequent generations.

Inevitably, human culture is built upon what came before. It is common to reflect on which ideas, innovations, and individuals contributed most to our current state of knowledge, but we don't often ponder how advancements that still lie to our future could have shaped certain aspects of our past. However, in the presence of a time machine capable of reaching more distant periods of the past, we must begin to consider how linkages through time could form fundamentally different relationships among people and events that exist as part of each integrated slice of 4-dimensional block time. A time machine developed at any point in our future, if capable of visiting us "now," would require that we too begin to form a broader conceptualization of a more complex system of intertemporal interconnectedness.

12.2. Intertemporal Integration

If backward time travel becomes a future reality, the intentional or unintentional exchange of information across the fourth dimension would need to be considered as part of our holistic existence. We would naturally be expected to gain more clarity regarding the nuances of this interaction, as we drew nearer to the time of the machine's creation. Though at any point within the temporal range of a device such as this, we cannot necessarily consider all cultural, technological, and perhaps even biological changes as purely sequential, given that if the arrow of time bends back upon itself, it ceases to function as an arrow, and instead becomes something of a knotted shoelace.

For example, imagine that the U.S. military actually did acquire a highly advanced craft at Roswell, New Mexico, in 1947, which was used by our distant descendants to return from the future to that particular point in our collective past. As physicists and engineers begin to dissect it and work to reverse engineer its components, they inevitably "create" the same knowledge that contributes to the technology necessary to develop the same time machine that is destined to smash into the New Mexico desert in 1947. These amalgamated moments—the craft's departure in the future, its crash in the past, as well as all related occasions in between—exist as an integrated network of self-consistent events, which all lie within the boundaries of this closed timelike curve (CTC).

Scenarios like this obscure the question of who is responsible for the actual creation of this technology. In other words, if knowledge of how to construct a backward time travel device is acquired by members of the past, and that information is then used to develop the same craft that will eventually result in its own creation, then who ultimately invented it? It certainly was not the past peoples who just happened upon this technology, and at a time that long-preceded its future existence. It also could not have been those in the distant future, given that their knowledge of this technology is largely dependent upon an evolutionary process that began when this futuristic device fell into the lap of their ancient ancestors.

Linkages like these across different regions of block time demonstrate how complexity mounts in instances of added intertemporal interaction. This is apparent in situations where materials from the future make their way into the past, though even the posterior transmission of information sent back across the iron curtain of time can be seen to rouse complexity. Admittedly, circumstances such as these can seem rather paradoxical and are jarring to our conventional notion of cause and effect. However, the majority of academic research still suggests that this added intricacy does not preclude backward time travel. In spite of these perceived incongruities, interconnectedness among events that lie along a CTC can still be seen to fit snugly within the warm comforting bosom of physical reality.

12.2.1. *Causal Loops and Self-consistency*

Perceived causality violations, in which something from the future aids in its own creation by being a part of its own past, is referred to as a *causal loop paradox*. This represents one of two broad categories of paradoxes that also includes the aforementioned *consistency paradox*.[463] The common heuristic device for understanding causal loops typically involves someone creating something or having an impactful idea, which, through the mechanism of backward time travel, is revealed to an earlier version of themselves or someone else in the past. This past person then uses the newly acquired information for some outcome and later in life, they become the one traveling backward through time to convey the same information to their younger self in the past, in a sort of infinite causal loop of creation and dissemination.[464]

As a more specific example, imagine that in the year 2095, a woman named Lucy enters an archaeological graduate program at a prestigious university. Shortly thereafter, Lucy is approached by a man named Fred, who hails from a time in the distant future that postdates the advent of backward time travel technology. During their 2095 encounter, Fred states that Lucy hired him to stop in the year 3003 to pick up a copy of her finished thesis to bring back to her in the year 2095. This dissertation centers on a lost civilization that she will soon "discover" that will go on to revolutionize our understanding of the origins of agriculture and the rise of civilizations throughout the world.

Over the course of the next eight years—largely based on information obtained from the thesis provided to her by her future self—she does indeed locate the lost civilization and proceeds to write about the ancient peoples who once occupied this exotic locale. After finishing work at the site, she compiles these data into a finished thesis, submits it to her graduate committee, and upon acceptance and publication, she proceeds to hire Fred to give her 2095-self a copy of the same dissertation, which will help revolutionize the anthropological subfield of archaeology, and so it goes.

Taken together, this scenario would seem to show an utter disregard for causality. For instance, if the dissertation exists at a time that predates its own creation, causality would appear to have been violated, at least in the context of our conventional notion of linear time. Additionally, the question of who actually discovered this ancient culture is seemingly quite paradoxical. However, when understood within the framework of a closed loop of self-consistent events, causality becomes irrelevant, and so does the question of who created the written content contained in the thesis.

In the same way that a rounded ball has no identifiable faces, edges, or vertices, one cannot pinpoint a specific origin of the thesis or who discovered the ancient civilization, as both the return-to and return-from points at each "end" of the loop exist as eternally integrated instances across that specific region of block time. The perplexity of these scenarios largely stems from the fact that our perception of time is habitually derived from obstinate observations of linearly organized events, with clearly defined boundaries between a past cause and a future effect. The question of who created content that has always existed without origin is perplexing, though it is also irrelevant in the context of relationships that exist among events that bridge different regions of four-dimensional spacetime.

Along with other distinguished colleagues of his time, Igor Novikov helped resolve many obscure facets of the causal loop paradox and contributed to a better understanding of consistency paradoxes as well. His work has also been instrumental in helping to resolve that nagging question of how something can be created by someone, and no one, at the same time. According to Novikov, the future may dictate events occurring in the past, because the future, present, and past are all one, and none are ahead of or behind the other in the presence of a time machine.[465] In this context, anything that has already happened will always have happened, and any future event that is to cause a past event—such as a PhD thesis creating itself—is not paradoxical, because it has also already taken place in both the future and past, regardless of where it is viewed in time.

Approaching these seemingly paradoxical situations in the context of self-consistent relationships among interconnected instances in block time makes it easier to conceptualize how they came to have always been. Intertemporal interactions such as these do not exist as capricious, self-determining occasions. Rather, they are forever interned as intricately linked bridges spanning different banks across the river of time. Considering the omnipresence of all events across all time, we must also acknowledge that it is impossible for anyone to ever *change* the past, simply because that past exists as part of a much larger and highly integrated whole, encompassing the entirety of spacetime. Additionally, to an observer who exists as part of the very last moment in the universe, it is, and has always been, the same long past, regardless of how many CTCs can be seen to weave their way in and out of the fabric of time.

12.2.2. Free Will and Self-consistency

It is common to think about different outcomes resulting from different decisions we feel we could have made at any time in our lives. For instance, "what would have happened if I had turned left at that stop light instead of right, would I still have been in that 30-car pileup?" "What would my life have been like if I was born to parents in Yemen instead of South Korea?" "What if I never had a horse, would I still have gone to college?" It is common to imagine alternative outcomes to any decision we feel we made. However, because of the nature of time, there actually aren't any. Self-consistency among events makes any perceived "change" to the past impossible, as any "alteration" was always part of the same inevitable outcome that already existed.

Another problem with the idea of changing things in our past is that if we alter an event that negatively impacted our future, then how would we ever know there was something we wanted to change in the first place, since it now never happened? In other words, if an unfortunate incident never occurred, then how could you know you wanted to return to the past to change it? Consider this example: you are part of a horrible auto accident. Later, you are approached by a benevolent time traveler from the future, who offers you the opportunity to go back and change anything about your life. You are likely to want to return to a time just before that accident, in order to avoid it. However, if you go back in time and convinced your younger self to alter the route you original followed, thus avoiding the crash, your knowledge of it, and your motivation for returning to that time to avoid it, suddenly disappears.

In spite of its unfeasibility, this scenario is commonly portrayed in film and television. Often, the outcome is that the time travel alters some aspect of the past and then returns home to find that everything is now different, and they are the only ones who are aware of it. This idea that each of us possesses the ability to change the future course of events is actually paradoxical. Alternatively, Novikov's self-consistency principle espouses that all events spread out across the landscape of time are intricately linked. As such, there can never be a "change" to the course of events that follow, given that any perceived alteration was always going to be a part of that which already existed.

With regard to the aforementioned auto accident example...if that crash occurred, it will always occur, because there cannot be two different series of events within the block universe. There is only one timeline for individuals, humans, this planet, the universe, etc. As much as we may regret the results of certain decisions we made, or circumstances that exist beyond our control, we often find that the outcome of a negative event inspired a positive one later on. Though regardless of how we perceive the valence of specific occurrences, either in real time, the immediate future, or in looking back upon the whole of our lives, it was always going to be the way it always was.

A good fictional representation of this self-evident self-consistency embedded in the essence of time can be found in Kurt Vonnegut's classic novel *The Slaughterhouse-Five*.[466] In this narrative, the principal protagonist named Billy Pilgrim is captured by a group of aliens called the Tralfamadorians. They steal Billy Pilgrim away from Earth and fly him to their distant planet, where he is meant to be kept as a zoo exhibit. The Tralfamadorians are a remarkable species, not least because they are capable of seeing all events that have ever been and that will ever be, all at the same time. As a result of this incredible ability, they are entertained by Billy Pilgrim's puerile concept of free will.

For instance, in noticing the Tralfamadorians' placating demeanor, Billy inquires as to how they managed to achieve such a peaceful coexistence, because he would like to take this knowledge back to Earth, to liberate his planet from the ravages of war. Since the Tralfamadorians are capable of seeing all time in both directions, they assure him that Earth will not be destroyed by war. Rather, they inform him that Earth will be annihilated when a Tralfamadorian test pilot—experimenting with new fuels for their flying saucers—presses the ignition button and causes the entire universe to disappear. As an earthly human whose mind is muddled by misguided notions of time and the illusion of free will, Billy Pilgrim asks "If you know this, isn't there some way you can prevent it? Can't you keep the pilot from pressing the button?" To which the Tralfamadorians reply "He has always pressed it, and he always will. We always let him and we always will let him. The moment is structured that way."[466]

If humans someday develop the materials, technology, and knowledge necessary to achieve backward time travel, we would also be expected to possess at least some knowledge of the future. Like the Tralfamadorians, we would be forced to recognize the certainty of certain predetermined moments in block time and, consciously or subconsciously, adhere to the self-consistent outcomes of these intertwined instants dispersed through-out it. We would undoubtedly continue to discuss aspects of daily life as if we have free will, even though we will continue to stay tied to the deterministic whipping post of time. What's more, we may also be forced to knowingly carry out some unpleasant tasks that we already know the outcome of, much like it was for Vonnegut's fictional Tralfamadorians.

A potential example of this may exist in the 1947 IFO crash near Roswell, New Mexico. Because this alleged event occurred in the past, future humans could potentially be aware of it. If they are knowledgeable of the outcome, this would require a conscious acceptance of the predetermined fate of the extratempestrials aboard, despite the unavoidable result of this deadly mission to the past. Furthermore, if they do possess knowledge of their looming destruction, it raises questions about whether they readily committed to carrying out what amounts to an intertemporal suicide mission, simply because the moment is structured that way.

It is also likely that these extratempestrials may not have known of their impending fate, for a couple of different reasons. Firstly, if enough time separates 1947 from whatever future period they originated from, historical records of the accident may have been lost. Secondly, due to psychological anxiety that is sure to be associated with knowing what is destined to happen to us—particularly regarding knowledge of when and how we are going to die—we may simply choose not to peer too deeply through time in either direction. In other words, we may choose to opt out of knowing our future, in an effort to maintain the biological and cognitive standards of reality that we have become accustomed to during our pre-time machine past.

It is fun to imagine what would happen if we traveled to the past to kill our grandfathers before they had our parents. Or if we returned to Africa 2.3-million-years-ago and wiped out every last member of our ancestral lineage, thus thwarting the existence of our entire species, in what would amount to a rather extreme version of the grandfather paradox. However, the fact that we exist now, means no one ever did and no one ever will.

Physics research has repeatedly shown that these and other perceived causality and consistency-paradoxes do not prohibit backward time travel, even though causal complexity mounts in association with increased intertemporal interaction. As stated previously, a desire to limit unnecessary convolution across different periods of human existence may be a component of why extratempestrials are not more overt in their activities and interactions with us. This is certainly not to say that an absence of evidence is evidence of their intentional absence. However, because of the complicated nature of backward time travel, this could conceivably be a critical component of why we have not yet seen more conspicuous occurrences of cross-temporal contact with extratempestrials.

12.3. Fermi's Paradox and Hawking's Time-Tourist Dilemma

As discussed earlier, physicist Enrico Fermi once famously asked: if there are billions of stars, with a certain percentage possessing earthlike planets, and a high probability of life arising on at least some of them, then where is everybody?[467] While he was focused on the question of extra*terr*estrial life and interstellar travel, the same question has also been

asked in the context of human time travelers from the future. More specifically, scholars have questioned why we are not inundated with tourists
from the future, if backward time travel is to someday become possible
for our distant descendants?

Physicist Stephen Hawking was among those who focused extensively
on this Fermiesque question, as it relates to the possibility of time travel
among members of the future human race. However, while Fermi was
open to the possibility of intelligent life on other planets, Hawking took a
much more critical approach to this time-tourist dilemma. Most notably,
he was fond of saying that because we are not overrun by future human
time tourists coming back to observe events of their past, this should be
taken as evidence that backward time travel will never be possible, at any
point in the human future.[468] Contrary to Hawking's objections, Paul
Davies—another well-known physicist and fellow time guru—takes a
more judicious position. In his book *About Time: Einstein's unfinished revolution*, Davies states:

> A shaky argument that is often used against time travel is that if
> our descendants ever discover how to do it they will come back
> and visit us. As we don't see these temponauts, we can conclude
> that they will never come to exist.[469]

More specifically, Davies counters Hawking's time-tourist dilemma argument by pointing out that some of the time machines that have been
modeled to date do not allow for backward travel beyond any point in
time prior to when the first time machine was invented. He argues that
unless our distant descendants use an element of nature to return to the
past, such as a wormhole for example, then we wouldn't expect to see
visitors from the future, at least until the first time machine is created.[470]

Davies' critique points to how a lack of evidence should not be considered evidence that a time machine will never exist. And while the overall message is important to consider, Davies' specific argument may not
actually be a limiting factor for our distant descendants, who may wish
to travel back in time beyond when the first time machine was created.
This is because there is not much support for the idea that the arrow of
time can only bend back upon itself up to the specific moment when a
backward time travel device was initially created. Bending light, space, and
time by means of a machine capable of producing enough energy to do
so, would likely not be limited by what humans had, or had not created,
at any particular point in their history. A machine cannot know when it is
supposed to stop moving backward through spacetime, and it would not

simply fade from existence once it passes the arbitrary point at which the first time machine was created, as this would be in strict violation of the law of conservation of mass.

As further evidence of Hawking's disdain for the notion that humans could ever achieve backward time travel, he also developed and was a staunch proponent of the aforementioned chronology protection conjecture. This advocates that the laws of physics are such that they will not allow backward time travel on all but a submicroscopic scale. It also states that the intrinsic properties of nature would prevent any alteration to the past that could result in the development of time paradoxes.

However, evidence suggests that there is only one past leading up to our present and, looking forward, there is only one future leading up to another person's present, as they look back on it as their own past. Within this framework, if it didn't already happen, it isn't going to happen, and nothing can be done to change that. So the idea that time-tourists from the future could somehow meddle with and change events of the past, runs counter to the bulk of the literature in this area, and most notably to the work of Novikov, Thorne, Friedman, Polchinski, Echeverria, Klinkhammer, Deutsch, Lloyd, Maccone, Ralph, Ringbauer, and others. This research has consistently shown that there are no initial conditions that lead to paradox in association with time travel to the past, particularly in association with nonrelativistic quantum mechanics, given that self-consistency remains globally consistent throughout. [464 471 472 473 474 475 476 477] As pointed out in a recent Scientific American article:

> Hawking and many other physicists find CTCs abhorrent, because any macroscopic object traveling through one would inevitably create paradoxes where cause and effect break down. In a model proposed by the theorist David Deutsch in 1991, however, the paradoxes created by CTCs could be avoided at the quantum scale because of the behavior of fundamental particles, which follow only the fuzzy rules of probability rather than strict determinism.[478]

Hawking was far less sanguine about the possibility of backward time travel, compared to many of his colleagues who believe the past requires no protection at all.[479] If self-consistency reigns, and an integrated network exists among events spread across the past and future as part of physical reality in the block universe, then what aspects of the past would require protection? If Novikov and his colleagues are correct, then the idea that the past could ever be changed is illogical, as the interjection of any information or individuals into that past were always a part of it. Additionally, and perhaps most importantly, if we can consider the multitude of reliable accounts of contact with extratempestrials as actual instances of intertemporal interaction, then these events would not only provide an answer to Hawking's chronology protection conjecture, but also to his Fermiesque question of "where is everybody?"

If reports of UFOs and "aliens" are indeed true, and these are travelers of deep time, rather than deep space, then clearly the laws of physics do not prohibit backward time travel. Furthermore, the presence of future peoples in our present and past would need to be considered as evidence for the existence of Davies' *temponauts*, and Hawking's *time-tourists*. In this context, an answer to the question 'why don't we see time travelers from the future if backward time travel is to someday become a reality,' may simply be, we do.

In spite of Stephen Hawking's strong belief in the incredible ability of the human race and the continued advancement of our species—which was echoed in his "message of hope" for the 1994 BT Group television advertisement—Hawking, somewhat inexplicably, fervently opposed the notion that we will ever develop the knowledge and technology necessary to achieve time travel to the past. With all due respect to one of the most resolute individuals and amazing minds of our time, this inconsistency is somewhat of a paradox unto itself.[480] To advocate that humans will never come to possess a deep enough understanding of time, nor the mechanisms with which to alter our position within it, is to deny millions of years of cerebral, cognitive, and cultural advancement across the collective whole of hominin evolution.

12.4. Temporal Gemination

Our quest to probe the upper atmosphere, space, the Moon, and Mars must have seemed absurd even as recently as 300 years ago, as we did not yet possess the prerequisite knowledge of physics, or the materials and machinery necessary to achieve such a feat. Although, considering the whole of human history, we actually reached our current state of technological advancement very quickly. Backward time travel is undoubtedly a more complex endeavor than sending someone to the moon. However, to advocate that we will never achieve this incredible feat sells short our entire species, as well as the long history of biocultural evolution that has brought us to where we are today.

There is still much we do not know about the elusive fourth dimension, as well as the logistics and latent effects of traveling backward through it.

One caveat that can cause consternation among those not accustomed to the anomalous nuances of backward time travel has to do with the potential for there to be two different copies of the same person or thing, in the same place, at the same time. This quandary centers on the perceived notion that if you go back in time, you are creating something that was never there in the first place. However, nothing inconsistent happens if someone returns to the past and meets a younger version of themselves, because that meeting always happened and always will, and it is remembered as such by both the younger and older versions of that same individual.

In other words, if you were to return to a moment within your own past, you were always a part of events that took place there and then, and you would always remember seeing your older self at that earlier time, right up until the point at which you become the older one returning to the past to meet your younger self. According to Caltech theoretical physicist Sean Carroll, in a 2010 *Discover Magazine* article titled *The Real Rules of Time Travelers*:

> If you met up with an older version of yourself, we know with absolute certainty that once you age into that older self, you will be there to meet your younger self. That is because, from your personal point of view, that meet-up happened, and there is no way to make it un-happen, any more than we can change the past without any time travel complications. There may be more than one consistent set of things that could happen at the various events in space-time, but one and only one set of things actually does occur.[481]

Those who reject backward time travel because of oddities involving interactions with past versions of the same people or things incorrectly impose a linear perception of time on one that is cyclic. When the future and past are linked together, events that are a part of one are naturally a part of the other as well. An odd outcome of this type of intertemporal interaction is that an encounter with your older self at a younger age provides a glimpse into your predestined future. More broadly, this aspect of time travel is analogous to what we may be experiencing with extratempestrial encounters. In other words, we may be offered a glimpse into the future biological and technological state of our species, as a result of meeting these older versions of ourselves during this earlier and less advanced stage of their evolutionary past. Carroll also addresses this aspect of time travel in his 2010 article *The Real Rules of Time Travelers*:

> A statement like 'We remember the past and not the future' makes perfect sense to us under ordinary circumstances. But in the presence of closed timelike curves, some events are in our past and also in our future....We know what is going to happen to us in the future because we witnessed it in our past.[481]

As discussed previously—and at length in Carroll's article as well—understanding the connection among different points in time has important implications regarding our notion of free will. More specifically, if we witness our future actions in the past, the implication is that we are predestined to carry out those same actions as part of our future, regardless of how freely we will ourselves to do otherwise. One important predetermined outcome is that we will continue to exist, at least up to the point that we become our future-self, who is fated to travel back to that same place and time, always and forever.

Many of the perplexing properties of backward time trave—such as the oddity of interacting with a past version of yourself—seem less problematic when considered in the context of future time travel. For instance, if one were to travel to the future to meet an *older* version of themselves, they are not likely to find them there, or then, given that their younger self disappeared from what would have been the world line leading up to that future existence, once they leapt forward in time. However, one should be able to meet a younger version of themselves in the past, since that individual was always present along the world line leading up to their departure backward in time. Although, this particular scenario—where there are now two versions of the same person in existence at the same time—does beg the question: has something been created from nothing in the process of returning to one's own past?

12.4.1. The Law of Conservation of Mass and Closed Time-like Curves

For those of us who move linearly through four-dimensional spacetime, a time traveler who appears out of nowhere creates the impression that something has been created from nothing. Furthermore, a sudden appearance such as this would seem to violate the law of conservation of mass, which states that matter can be neither created nor destroyed, and that the mass of a system must remain constant over time. However, time travel does not actually create or destroy any matter, nor does it violate any physical laws, even at the level of quantum particles as shown previously. Rather, this is simply a rearrangement of existing matter in and around different regions of the block universe.

Moving people and things through spacetime by means of a high-energy device capable of forming CTCs merely changes *when* things are relative to one another. So, despite the perception that something is suddenly popping into existence, this aspect of time travel is better understood as a mere relocation of existing matter, within a much larger system that extends well beyond the present. As such, this perceived paradox may also stem from our tendency to regard time as linear and the present as a closed system, as opposed to one that encompasses the entire universe and the whole of four-dimensional spacetime.

Misapprehension regarding this aspect of backward time travel is a bit like how we view our surroundings prior to the development of object permanence. For instance, if you place an apple in front of a 4-month-old baby while it is looking away, once they notice it, the baby may think you are some sort of sorcerer capable of conjuring apples at will. If you had taken that particular apple away from a different 4-month old baby in the process, that baby may assume the apple disappeared from existence altogether. As a cognizant adult, it is clear that you did not magically create or destroy any apple-matter, but simply moved an existing apple across three-dimensional space, relative to the field of vision of these two babies.

Moving an apple between two babies occupying different periods of time is quite similar. If you take the apple from future-baby and deliver it to past-baby, it would again appear to vanish, though the apple itself still exists, just in a region of spacetime that is now outside future-baby's frame of reference. A similar scenario unfolds for past-baby on the other end of the time loop. They observe the apple suddenly popping into existence, despite the fact that it had only moved from future to past, while maintaining its same position in space and without ever being created or destroyed in the process.

Once we establish the permanence of objects within our shared region of space and time, we can quickly deduce that if something is suddenly gone, it must have moved elsewhere while continuing to exist as part of this thing we call the present. Though if something suddenly disappears from plain sight, we are not conditioned to consider that it might still exist in the same place, but at a time that is no longer a part of our collective present. This is likely a part of why time travel is so bewildering to us. During the earliest stages of life, we work very hard to understand *where* things go, but we have not yet been forced to consider *when* things go as well.

It is also important to consider that when a person, an apple, or anything else goes back in time, as mentioned above, it ceases to persist into the future, following the time from which it left for the past. This represents something of a four-dimensional zero-sum game. In other words, if we enter a spinning disc and go back in time to meet an earlier version of ourselves—despite their now being two copies of us in the past—there are no longer any copies of us directly following when we left to visit ourselves in that past. So even though there are now two versions of the same person existing concurrently in the past $(1 + 1 = 2)$, there are no longer any versions of that individual in the future after they left for the past $(1 - 1 = 0)$, and when averaged across the entire system, the result is still just one individual.

12.5. Temponauts, Time-Tourists, and Extratempestrials

Accounts of direct contact with extratempestrial "aliens" who reportedly possess the same, yet slightly more derived, morphological and cultural characteristics of past and modern hominins, suggests we may be destined to develop the materials, technology, and knowledge necessary to achieve backward time travel. While our time traveling descendants may potentially be understood in the context of Paul Davies' *temponauts*, it is not likely that these individuals are merely *time-tourists*, as considered by Stephen Hawking. To the contrary, based on reliable reports describing what these extratempestrials do while in our time—and how similar these actions are to what we anthropologists would do if capable of studying our own ancestors with the same technology—it seems more probable that these are scientist engaged in investigative research.

With that said, it is certainly possible that some individuals aboard these IFOs are in fact tourists, who have no doubt paid a handsome sum for the luxury of touring a past period of time. Time tourism would surely be a highly lucrative venture for anyone capable of bringing tourists back from the future to observe the primitive ways past peoples lived, how the great monuments were built, or to observe plants and animals that went extinct long before their time. Indeed, a similar type of time-tourism already exists, as we often seek out and spend large sums of money for the chance to visit people and places that appear frozen in time.

It is common for present peoples to travel long distances and at great expense to observe the ancient relics of Machu Picchu, The Great Wall of China, the Pyramids of Egypt, the homesteads of the Amish, or to watch traditional peoples perform traditional dances in traditional garb, only to return to jeans and T-shirts as soon as the tourists stumble off to bed. In fact, according to a 2015 TripAdvisor report, 19 of the 25 most popular tourist attractions in the world are historic or prehistoric sites, with Angkor Wat, Cambodia, and Machu Picchu, Peru, sitting at the top of the list.[482]

If we are someday able to achieve backward time travel, there is certain to be a long line of people who would jump at the opportunity to witness firsthand the endless plethora of important, and even tremendously mundane moments that have shaped human history on this planet. In fact, heavy demand for time tourism may even help fuel progress in the development of a time travel device. After all, it is abundantly clear from researching human economic activity across space and through time, that if there is money to be made from doing something, someone will find a way to do it.

In spite of what would surely be strong demand among future humans for the chance to partake in taking in the past, some legal restrictions may be placed on time travel due to complexities that result from linking together different points in time. As stated previously, modern physics research suggests that intertemporal interaction does not preclude backward time travel, though it undoubtedly complicates the path of individual world lines meandering through block time. So, in order to limit visibility—and surely outright contact between future and past peoples—multilateral legislative limits are likely to be put in place ahead of any mission to the past, and especially to periods that predate widespread public knowledge of time travel technology.

As stated earlier in the text, this issue of added time-loop ambiguity may help account for why we are not yet overrun by time-tourists, as well as why most reported experiences with extratempestrials imply that we are simply research subjects in a holistic study of human biocultural evolution. In effect, systematic scholarly studies of the past may be the only permissible pursuit deemed worthy of risking unintended intertemporal interjection and chronological convolution. In other words, if complexity mounts as discrete time periods become intertwined, future humans may be incentivized to avoid becoming an already-existing, overly overt component of their own past.

With this in mind, it is perhaps not a coincidence that the modus operandi of extratempestrials seems to mimic the covert operations of a stealthy Navy Seals Unit, or a snow leopard stalking a blue sheep in the Himalayas of Tibet. This is further indicated by the tremendous number of credible reports of close encounters in which individuals are approached at odd hours of the night and in very remote regions of the world. Such behavior would be expected of extratempestrials attempting to avoid implicitly interjecting themselves into past periods, particularly those that predate broad knowledge of their eventual existence.

The caveat of chronological complexity may be a principal reason why we are not yet inundated by time-tourists, everywhere and everywhen, despite indications that we will eventually become capable of visiting the past. In the very least, aspects of intertemporal obscurity could help explain why extratempestrials seem so elusive in their pursuits, as strict limits may be placed upon scientists, tourists, party-goers, equities trad-ers, sports gamblers, or anyone else who wishes to travel through time, mandating that any visit to the past remains as surreptitious as possible, regardless of the intent.

12.6. Eventual, Intentional, Intertemporal Interaction

Regardless of whether they are scientists, tourists, pervy voyeurs, pro-leptic proctologists, or time-criminals bent on stealing the riches of the past, the multiple-thousands of legitimate reports of close encounters strongly suggests that time travel will not only be possible, but that we will be the ones doing it. As noted previously, extraterrestrial space aliens who traverse vast stretches of the universe to visit Earth would be expected to at least get out and say hello. However, our future descendants, traveling backward through time to visit us in their own past, would not. If close encounters are in fact real events, and if we are inclined to choose between an extraterrestrial or an extratempestrial model to explain the origin of those involved, an abductive approach—considering all of the available evidence—points to a time travel, rather than a space travel explanation of this phenomenon.

There are simply too many flaws with the idea that human-like beings could ever evolve on a different planet in a distant solar system, that they could find us here among all of the other stars in the universe, or that they could, or ever would traverse exceptionally long distances simply to harass primitive Earthlings in utter secrecy over long periods of time. To the contrary, it is much easier to understand why time-traveling humans from the future would look like us, how they would find us, how they would get here, and why they would refrain from ostentatious encounters with past persons, or at least until their existence became widely appreciated as such.

In fact, many perceived paradoxes associated with cross-temporal con-tact may only matter during periods when secrecy regarding these future abilities is stalwartly maintained. Though at some point in the future, if time travel is to become a part of our conventional reality, encountering someone from a different time may become as common as encountering someone from a different continent, which itself was nearly nonexistent even 600 years ago. Once these barriers to anachronous activity are broken down, humans from distinct periods throughout the post-disclosure future could choose to weave a much more complex tapestry from the fabric of time, without the cognitive hang-ups of contemporary culture.

Unfettered knowledge of our future ability to time travel could facilitate a far more holistic and scientific study of humankind, and the opportunity to investigate the intricacies of intertemporal interaction across broad swathes of our existence on this planet. For now, the deep past and even the immediate future lock away their secrets, but with sustained progress in developing a deeper understanding of time and the human condition, the darkness of the future may begin to brighten, revealing a brilliant utopia of knowledge radiating outward in all directions across the omnipresent landscape of time.

This expanded state of consciousness is also likely to precede any real ability to achieve backward time travel. For if we are destined to eventually become the extratempestrials observed now and in the past, we would expect to learn of this imminent reality long before the creation of any actual time machine, at whatever point in the future our time traveling descendants wished to inform us of it. Furthermore, in learning of—and certainly in becoming our cross-temporal progeny—we would usher in a new phase of human history, in which individuals and events exist as overtly integrated entities across disparate periods, forming a more reticulated network of predetermined causality, among an increasingly progressive and pervasive intertemporal human race.

Endnotes

Epigraph

1   Wheeler, John Archibald - Renowned theoretical physicist. Quoted in Gleick, James (1993) *Genius: The Life and Science of Richard Feynman*. New York: Pantheon Books.

Chapter 1

2   Lederman, Leon M. (1993) - Renowned physicist and author of *The God Particle: If the Universe Is the Answer, What is the Question?* (pp. 175). Boston, MA (USA): Houghton Mifflin Harcourt.

3   Wilde, Oscar. (2007) *The Collected Works of Oscar Wilde* (p.1001). Ware, Hertfordshire (UK): Wordsworth Editions Limited.

4   Strieber, W. (1987) *Communion: A True Story.* Sag Harbor, New York (USA): Beech Tree Books.

5   Image Credit: Mad Dog/shutterstock.com (left), Makc/shutterstock.com (center), Brian Goff/shutterstock.com (right)

6   Term coined by Dr. Robert Martin, Curator Emeritus at the Chicago Field Museum. Personal communication, 2015.

7   Cross, A. (2004) The flexibility of scientific rhetoric: A case study of UFO researchers. *Qualitative Sociology*, 27(1), 3-34.

8   Darwin, C., & Wallace, A. (1858) On the tendency of species to form varieties; and on the perpetuation of varieties and species by natural means of selection. *Journal of the Proceedings of the Linnean Society of London. Zoology*, 3(9), 45-62.

9   See Kean, L. (2011) *UFOs: Generals, Pilots, and Government Officials Go on the Record.* Three Rivers Press (CA).

10   Hynek, J. A. (1972) *The UFO experience: A scientific inquiry*. Chicago, IL (USA): Henry Regnery.

11   See Brookesmith, P. (1995) *UFO: The complete sightings*. New York, NY (USA): Barnes & Noble.

12   See Randle, K. D. (1995) *A history of UFO crashes*. New York, NY (USA): Avon Books.

13   Cooper, A. (2007, March 21) Former governor says he saw UFO. [Web log post] Retrieved August 19, 2015 from: http://www.cnn.com/CNN/Programs/anderson.cooper.360/blog/2007/03/former-governor-says-he-saw-ufo.html

14   Huffington Post and YouGov poll of 1,000 adult individuals asking if they either believed or didn't believe that some people have witnessed UFOs

that have an extraterrestrial origin. The poll was conducted September 6-7, 2013 and included individuals of diverse age, political, ethnic, educational, gender, and religious backgrounds. Results retrieved September 11, 2015 from: http://big.assets.huffingtonpost.com/tabs_ufo_0906072013.pdf

15   Quoted in Speigel, L. (2013, September 11) 48 Percent of Americans Believe UFOs Could Be ET Visitations. *Huffington Post*. Retrieved September 11, 2015 from: http://www.huffingtonpost.com/2013/09/11/48-percent-of-americans-believe-in-ufos_n_3900669.html

Chapter 2

16   Fife Symington III – Former Governor of Arizona, U.S. and witness to an I.F.O event, speaking during an interview on the television program *Secret Access: UFOs on the Record*. Originally aired Thursday, August 25, 2011 at 8 p.m. ET on the History Channel, based on *The New York Times* bestseller by Leslie Kean (2011) *UFOs: Generals, Pilots, and Government Officials Go on the Record*. Three Rivers Press (CA).

17   Nick Pope, UK Ministry of Defense, 1985 – 2006, speaking during an interview on the television program *Secret Access: UFOs on the Record*. Originally aired Thursday, August 25, 2011 at 8 p.m. ET on the History Channel, based on *The New York Times* bestseller by Leslie Kean (2011) *UFOs: Generals, Pilots, and Government Officials Go on the Record*. Three Rivers Press (CA).

18   Richard Haines, Ph.D. – Chief Scientist, National Aviation Reporting Center on Anomalous Phenomena (NARCAP), speaking during an interview on the television program Secret Access: UFOs on the Record. Originally aired Thursday, August 25, 2011 at 8 p.m. ET on the History Channel, based on *The New York Times* bestseller by Leslie Kean (2011) *UFOs: Generals, Pilots, and Government Officials Go on the Record*. Three Rivers Press (CA).

19   Liverpool, J. D., BT and Saatchi & Saatchi advertising agency (1995) *Inspiring British Telecom TV ad featuring Stephen Hawking 1995* [Television Commercial]. Retrieved September 11, 2015 from: http://toptvadverts.com/inspiring-british-telecom-tv-ad-featuring-stephen-hawking-1995/

20   Gilmoure, D., Wright, R., Samson, P. (Released 1994, March 12) Keep Talking [Recorded by Pink Floyd] On *The Division Bell*. London, United Kingdom: Columbia Records.

21   Michaels, S. (2014, October 8) Stephen Hawking sampled on Pink Floyd's The Endless River. *The Guardian*. Retrieved September 11, 2015 from: http://pfco.neptunepinkfloyd.co.uk/band/interviews/grp/grpredbeard.html.

22   Bickerton, D. (1981). *Roots of language*. Karoma.

23   Bickerton, D. (2007). Language evolution: A brief guide for linguists. *Lingua, 117*(3), 510-526.

24   Welles, Orson (1938, October 30) The War of the Worlds. In Howard Koch (Writer), *Mercury Theater on the Air*. CBS Radio. WCBS, New York.

     Note. War of the Worlds. Adapted from H. G. Wells (1897) *War of the Worlds*

25   Lovgen, S. (2005, June 17) War of the Worlds: Behind the 1938 Radio Show Panic. *National Geographic News*. Retrieved November 12, 2015 from http://news.nationalgeographic.com/news/2005/06/0617_050617_warworlds.html

26   Sample, I. (2010, January 24) Alien visitors to Earth may be as acquisitive as humans. *The Guardian*. Retrieved August 14, 2015 from: http://www.theguardian.com/science/2010/jan/25/aliens-space-earth-humans

27   Gleiser, M. (2015, February 11) Should We Be Afraid Of Aliens? *Montana Public Radio: 13.7 Cosmos and Culture NPR*. Retrieved February 25, 2015 from: http://www.npr.org/blogs/13.7/2015/02/11/385413799/should-we-be-afraid-of-aliens

28   National Archives. *Unidentified Flying Objects—Project BLUE BOOK*. Retrieved November 2014 from: http://www.archives.gov/research/military/air-force/ufos.html

29   BBC News UK (2013, June 20) UFO Sightings: Files Explain Why MoD Closed Down Special Desk. *BBC News UK*. Retrieved November 2014 from: http://www.bbc.com/news/uk-22991014

30   Smetanina, S. (2013, April 12) Former KGB Agent Reveals Soviet UFO Studies. *Russia Beyond the Headlines*. Retrieved on November 2014 from: http://rbth.com/science_and_tech/2013/04/12/former_kgb_agent_reveals_soviet_ufo_studies_24927.html

31   Losey, S. (January 20, 2015) Air Force UFO files hit the web. *Military Times*. Retrieved June, 2015 from http://www.militarytimes.com/story/military/tech/2015/01/17/air-force-ufo-files/21812539/

32   Haines, G. K. A Die Hard Issue: CIAs Role in the Study of UFOs 1947-90. Retrieved November 2014 from National Investigations Committee on Aerial Phenomena official web site: http://www.nicap.org/ciarole.htm

33   Hynek, J. A. (1972) *The UFO Experience: A scientific Inquiry*. Chicago, IL. Henry Regnery Company.

34   Vallée, J. (1998) Physical Analysis in Ten Cases of Unexplained Aerial Objects with Material Samples. *Journal of Scientific Exploration* 12(3), 359-375.

35   McCarthy P. (1992) Close encounters of the fifth kind - Communicating with UFOs. *OMNI Magazine*.

36   REPORT OF SCIENTIFIC ADVISORY PANEL ON UNIDENTIFIED FLYING OBJECTS CONVENED BY OFFICE OF SCIENTIFIC INTELLIGENCE, CIA (January 14 - 18, 1953) The Durant report

of the Robertson Panel proceedings. Retrieved November 2014 from: http://www.cufon.org/cufon/robert.htm

37   Library and Archives Canada. Canada's UFOs: The Search for the Un-known. *Library and Archives Canada.* Retrieved September 11, 2015 from: http://www.collectionscanada.gc.ca/ufo/002029-1400.01-e.html

38   Smith, W. B. (1950, November 21) GEO-MAGNETICS, Department of Transport. Cited in Library and Archives Canada. Canada's UFOs: The Search for the Unknown. *Library and Archives Canada.* Retrieved September 11, 2015 from: http://www.collectionscanada.gc.ca/ufo/002029-1400.01-e.html

39   Smith, W. B. (1952) Box 7523. File DRBS 3800-10-1, pt. 1. Project Magnet Report (pp. 6). Library and Archives Canada, Ottawa, Ontario, Canada. Retrieved November 17, 2015 from: http://www.collectionscanada.gc.ca/ufo/002029-1401-e.html

40   Huneeus, J. A. (1993, August) Beyond the Belgian Flap—UFOs and Euro-politics. FATE Magazine. Retrieved October 2014 from: http://www.ufoevidence.org/documents/doc405.htm#FairUse

41   For an in-depth analysis of the psychological aspects of this phenomenon see Jung, C. G. (2014) *Flying saucers: A modern myth of things seen in the sky.* Psychology Press.

42   Novella, S. (2000, October) UFOs: The Psychocultural Hypothesis. *The New England Skeptical Society.* Retrieved June 1, 2018 from: https://theness.com/index.php/ufos-the-psychocultural-hypothesis/

43   Feltman, R. (2015, July 20) Stephen Hawking announces $100 million hunt for alien life. *The Washington Post: Speaking of Science.* Retrieved August 20, 2015 from: http://www.washingtonpost.com/news/speaking-of-science/wp/2015/07/20/stephen-hawking-announces-100-million-hunt-for-alien-life/

44   Liverpool, J. D., BT and Saatchi & Saatchi advertising agency (1995) *Inspiring British Telecom TV ad featuring Stephen Hawking 1995* [Television Commercial]. Retrieved September 11, 2015 from: http://toptvadverts.com/inspiring-british-telecom-tv-ad-featuring-stephen-hawking-1995/

Chapter 3

45   Sagan, C. (1979). *Broca's Brain, Reflections on the Romance of Science* (pp. xiv). New York: Random House.

46   Shaw, I. (2002). Building the Great Pyramid. *BBC History.*

47   Stocks, D. A. (2013). *Experiments in Egyptian archaeology: stoneworking technology in ancient Egypt.* Routledge.

48   Coles, J. (2014). *Archaeology by experiment* (Vol. 16). Routledge.

49   Däniken E. V. (1972) *Chariots of the Gods.* Berkley Publishing Group. New York, NY.

50   Enlow, D. H., & Hans, M. G. (Eds.). (1996). *Essentials of facial growth.* WB Saunders Company.

51   Lieberman, D. E., Ross, C. F., & Ravosa, M. J. (2000). The primate cranial base: ontogeny, function, and integration. *American Journal of Physical Anthropology, 113*(s 31), 117-169.

52   Lieberman, D. E., Pearson, O. M., & Mowbray, K. M. (2000). Basicranial influence on overall cranial shape. *Journal of Human Evolution, 38*(2), 291-315.

53   Bastir, M., Rosas, A., & O'Higgins, P. (2006). Craniofacial levels and the morphological maturation of the human skull. *Journal of Anatomy, 209*(5), 637-654.

54   Bastir, M., O'Higgins, P., & Rosas, A. (2007). Facial ontogeny in Neanderthals and modern humans. *Proceedings of the Royal Society of London B: Biological Sciences, 274*(1614), 1125-1132.

55   Bastir, M., Rosas, A., Lieberman, D. E., & O'Higgins, P. (2008). Middle cranial fossa anatomy and the origin of modern humans. *The anatomical record, 291*(2), 130-140.

56   Cheverud, J. M., Kohn, L. A., Konigsberg, L. W., Leigh, S. (1992) The effects of fronto-occipital artificial cranial vault modification on the cranial base and face. *American Journal of Physical Anthropology* 88:323-346.

57   Cheverud J. M., Midkiff, J. (1992) The effects of fronto-occipital cranial reshaping on mandibular form. *American Journal of Physical Anthropology* 87,167-172.

58   Trinkaus E. (1982) Artificial cranial deformation in the Shanidar 1 and 5 Neanderthals. *Current Anthropology 23*,198–199.

59   Smith, G. E. (1932) Artificial cranial deformation: A contribution to the study of ethnic mutilations. *Nature, 130*, 185-186.

60   Gerszten P. C, Gerszten E. (1995) Intentional cranial deformation: A disappearing form of self-mutilation. *Neurosurgery, 37*(3)374-382.

61   Tubbs R. S., Salter E. G., Oakes W. J. (2006) Artificial deformation of the human skull: a review. *Clinical Anatomy, 19*(4)372-377.

62   Tiesler, V. (1999, March). Head shaping and dental decoration among the ancient Maya: Archaeological and cultural aspects. In *Proceedings of the 64th Meeting of the Society of American Archaeology* 24-28.

63   See for instance the controversial claim that *Homo naledi* (itself a somewhat controversial species designation) was intentionally burying its dead, possibly as early as 1 – 1.5 million years ago: Dirks, P. H., Berger, L. R., Roberts, E. M., Kramers, J. D., Hawks, J., Randolph-Quinney, P. S., ... & Schmid, P. (2015) Geological and taphonomic context for the new hominin species *Homo naledi* from the Dinaledi Chamber, South Africa. *eLife, 4*, e09561.

64   See for example the recent media criticisms by Berkeley Professor of Biological anthropology Tim White, regarding the designation of *Homo naledi*

as a new species. As well as criticisms of this new form of media-based criticism in Martin, G. (October 1, 2015) Bones of Contention: Why Cal Paleo Expert is So Skeptical That *Homo Naledi* Is New Species. *Cal Alumni Association*, UC Berkeley. Retrieved October 9, 2015 from: http://alumni.berkeley.edu/california-magazine/just-in/2015-10-05/bones-contention-why-cal-paleo-expert-so-skeptical-homo

65 This is a common mantra among skeptics, and while often attributed to Sagan, it may actually predate him. See Farley, T. (November 4, 2014) A Skeptical Maxim (May) Turn 75 This Week. Skeptic. Retrieved October 9, 2015 from: http://www.skeptic.com/insight/open-mind-brains-fall-out-maxim-adage-aphorism/

66 Sagan, C. (1990). Why we need to understand science. *Skeptical Inquirer, 14*(3), 263-9.

67 Iwase Bunko Library was established as a private library in Nishio, Japan in 1908 by the wealthy merchant Yasuke Iwase. Retrieved September 23, 2015 from: http://www.iwasebunko.com/

68 Tanaka, K. (2000) Did a Close Encounter of the Third Kind Occur on a Japanese Beach in 1803?. *Skeptical Inquirer, 24*(4), 37-41.

69 Sagan, C. (2011) *Demon-haunted world: Science as a candle in the dark*. New York City, New York: Ballantine Books.

70 Randle, K. D. and Schmitt, D. R. (1994) *The Truth about the UFO Crash at Roswell*. Lanham, MD: M Evans & Co.

71 Pflock, K. T. (2001) *Roswell: Inconvenient Facts and the Will to Believe*. Amherst, NY: Prometheus Books.

72 Wilson, J. (1997, July) Roswell Plus 50. Roswell – Starling New Revelations About The UFO Crash Coverup 50 Years Ago. *Popular Mechanics, 174(7)*, 48-53.

73 Report of Scientific Advisory Panel on Unidentified Flying Objects Convened by Office of Scientific Intelligence, CIA (1953, January 14-18) *The Durant Report of the Robertson Panel*. Retrieved October 21, 2015 from: http://www.cufon.org/cufon/robert.htm

74 Clark, J. (2003) *Strange Skies: Pilot Encounters with UFOs*. New York, New York: Citadel Press.

75 Clark, J. (1998) *The Lubbock Lights*, from *The UFO Book* (pp. 342-350). Detroit: Visible Ink Press.

76 Ruppelt, E. J. (1956) *The Report on Unidentified Flying Objects* (pp. 96-110). New York: Doubleday.

77 Keyhoe, D. E., USMC Maj. (ret.) (1955) *The Flying Saucer Conspiracy* (pp. 13-15). New York, New York: Henry Holt & Co.

78 For full report see Department of the Air Force Headquarters Safety Agency (n.d.) *Kinross AFB Missing F-89C – 23 Nov. 1953*. Retrieved October 21, 2015 from: http://cufon.org/kinross/Kinross_acc_rept.htm

79   Appelle, S. (1996) The Abduction Experience: A Critical Evaluation of Theory and Evidence. *Journal of UFO Studies*, 6, 29–78.

80   Webb, W. (1961) A Dramatic UFO Encounter in the White Mountains, New Hampshire: The Hill Case—Sept. 19-20, 1961. *Confidential NICAP Report, October, 26.* Retrieved October 23, 2015 from: http://www.nicap.org/reports/610919hill_report2.pdf

81   Friedman, Stanton & Kathleen Marden (2007) *Captured! The Betty and Barney Hill UFO Experience.* Franklin Lakes, NJ: New Page Books.

82   Friedman, S. T, Marden, K. (2007) *Captured! The Betty and Barney Hill UFO Experience.* Franklin Lakes, NJ: New Page Books.

83   The National UFO Reporting Center (NUFORC) database. Retrieved July 25, 2014 from: http://www.nuforc.org

84   Data obtained from the National UFO Reporting Center (NUFORC) database. (2014, July) Monthly Report Index for 07/14. Retrieved July 25, 2014 from: http://www.nuforc.org/webreports/ndxe201407.html

85   UFOlogy Research of Manitoba. The Canadian UFO Survey. Retrieved July 25, 2014 from: http://www.canadianufosurvey.com

86   Data compiled from database query. Retrieved July 25, 2014 from: http://www.canadianufosurvey.com/Search

87   Retrieved July 25, 2014 from: http://www.ufocenter.com/reportform.html

88   Kiger, P. J. (2012) Top 10 Mass Sightings of UFOs. *National Geographic.* Retrieved December 10, 2015 from: http://channel.nationalgeographic.com/chasing-ufos/articles/top-10-mass-sightings-of-ufos/

89   NOAA, National Oceanic and Atmospheric Administration, United States Department of Commerce. Retrieved December 15, 2015 from: http://www.noaa.gov/features/02_monitoring/balloon.html

Chapter 4

90   Carl Sagan (1978) The Quest for Extraterrestrial Intelligence. *Cosmic Search Magazine, 1*(2). Retrieved April, 2015 from: http://www.bigear.org/vol1no2/sagan.htm.

91   Jones, E. M. (1985) Where is everybody? An account of Fermi's question. *NASA STI/Recon Technical Report N, 85,* 30988.

92   Drake, F., & Sobel, D. (1992) *Is anyone out there?.* Delacorte Press.

93   Tipler, F. J. (1980) Extraterrestrial intelligent beings do not exist. *Quarterly Journal of the Royal Astronomical Society, 21,* 267-281.

94   Leopold, T. (2015, May 14) NASA Chief Scientist: "Indications" of alien life by 2015. *CNN.* Retrieved November 24, 2015 from: http://www.cnn.com/2015/04/08/us/feat-nasa-scientist-alien-life/

95   HarvardX: SPU30x Super-Earths and Life (2015) EdX online course in Astrobiology through Harvard University's HarvardX program.

96   Dick, S. J. (2006) Anthropology and the search for extraterrestrial intelligence: An historical view. *Anthropology Today, 22*(2), 3-7.

97   Vakoch, D. A. (2009) Anthropological Contributions to the Search for Extraterrestrial Intelligence. *Bioastronomy 2007: Molecules, Microbes and Extraterrestrial Life, 420,* pp. 421.

98   See especially National Aeronautics and Space Administration (2014). *Archaeology, Anthropology, and Interstellar Communication.* Vakoch D. A. (Ed.). Office of Communications, Public Outreach Division.

99   Drake, F.D. Discussion at Space Science Board-National Academy of Sciences Conference on Extraterrestrial Intelligent Life, November 1961, Green Bank, West Virginia.

100 Drake, F. (1965). The radio search for intelligent extraterrestrial life. *Current aspects of exobiology*, 323-345.

101 The Drake Equation. Retrieved August, 2014 from: http://www.foothill.edu/attach/938/Drake_equation.pdf

102 Cirkovic, M. M. (2004) The temporal aspect of the Drake equation and SETI. *Astrobiology, 4*(2), 225-231.

103 Clavin, W. (2015, October 6) Exoplanet Anniversary: From Zero to Thousands in 20 Years. NASA Jet Propulsion Laboratory, Pasadena, California. Retrieved December 17, 2015 from: http://www.jpl.nasa.gov/news/news.php?feature=4733&linkId=17750636.

104 Chick, G. (2014) Biocultural Prerequisites for the Development of Interstellar Communication. *Archaeology, Anthropology, and Interstellar Communication* (pp. 215). Vakoch, D. A. (Ed.). National Aeronautics and Space Administration. Office of Communications, Public Outreach Division.

105 Vakoch, D. (2014) Reconstructing Distant Civilizations and Encountering Alien Cultures. *Archaeology, Anthropology, and Interstellar Communication* (pp. xiv). Vakoch, D. A. (Ed.). National Aeronautics and Space Administration. Office of Communications, Public Outreach Division.

106 Harmand, S., Lewis, J. E., Feibel, C. S., Lepre, C. J., Prat, S., Lenoble, A., ... & Roche, H. (2015) 3.3-million-year-old stone tools from Lomekwi 3, West Turkana, Kenya. *Nature, 521*(7552), 310-315.

107 Latimer, B. (2005) Editorial: The Perils of Being Bipedal. *Annals of biomedical engineering, 33*(1), 3-6.

108 Latimer, B. (2005) Editorial: The Perils of Being Bipedal. *Annals of biomedical engineering, 33*(1), 3.

109 Data obtained from the PHL Exoplanet Catalog of the Planetary Habitability Laboratory at the University of Puerto Rico at Arecibo. CSV Database File: Confirmed Exoplanets: phl_hec_all_confirmed.csv. Retrieved January 7, 2016 from: http://phl.upr.edu/projects/habitable-exoplanets-catalog/data/database

110 Morey-Holton, E. R. (2003) The impact of gravity on life. *Evolution on planet earth: the impact of the physical environment*, 143-159.

111 Newman, D. J., Hoffman, J., Bethke, K., Carr, C., Jordan, N., Sim, L., ... & Trotti, G. (2005) An Astronaut 'Bio-Suit'System for Exploration Missions.

112 Williams, N. (Director). (2014, January 8) Alien Planets Revealed [Television series episode]. In Apsel, P. S. (Producer), *NOVA*. Boston, MA: WGBH. Retrieved June 6, 2015 from: http://www.pbs.org/wgbh/nova/space/alien-planets-revealed.html

113 Borucki, W. J., Agol, E., Fressin, F., Kaltenegger, L., Rowe, J., Isaacson, H., ... & Fabrycky, D. (2013). Kepler-62: a five-planet system with planets of 1.4 and 1.6 Earth radii in the habitable zone. *Science, 340*(6132), 587-590.

114 Countryman, S. M., Stumpe, M. C., Crow, S. P., Adler, F. R., Greene, M. J., Vonshak, M., & Gordon, D. M. (2015) Collective search by ants in microgravity. *Frontiers in Ecology and Evolution, 3*, 25.

115 Webb, J. (2015, March 31) Ants in space grapple well with zero-g. *BBC Science and Technology*. Retrieved May, 2015 From: http://www.bbc.com/news/science-environment-32115413

116 Morris, S. C. (2015) *The Runes of Evolution: How the Universe became Self-Aware.* West Conshohocken, PA: Templeton Foundation Press.

117 Human-like aliens are likely to have evolved on other planets – just as we did on Earth (2015, July 02) *The Irish Examiner*. Retrieved August 14, 2015 from: http://www.irishexaminer.com/examviral/science-world/human-like-aliens-are-likely-to-have-evolved-on-other-planets-just-as-we-did-on-earth-340294.html

118 Simpson, F. (2016) The size distribution of inhabited planets. *Monthly Notices of the Royal Astronomical Society: Letters, 456*(1), L59-L63.118

119 Behroozi, P., & Peeples, M. S. (2015) On the history and future cosmic planet formation. *Monthly Notices of the Royal Astronomical Society, 454*(2), 1811-1817.

120 Tipler, F. J. (1980) Extraterrestrial intelligent beings do not exist. *Quarterly Journal of the Royal Astronomical Society, 21*, 267.

121 Lineweaver, C. H., & Davis, T. M. (2002) Does the rapid appearance of life on Earth suggest that life is common in the universe?. *Astrobiology, 2*(3), 293-304.

Chapter 5

122 Corey S. Powell, Editor at large for Discover Magazine and Consulting Editor at American Scientist, demonstrating the tremendous distance between planets and stars in the article Where's My Warp Drive? (2015, May 28) *Discover Magazine*. Retrieved September 14, 2015 from: http://discovermagazine.com/2015/july-aug/31-wheres-my-warp-drive

123 Einstein, A. (1905) On the electrodynamics of moving bodies. *Annalen der Physik, 17*(891), 50.

124 Atri, D., DeMarines, J., & Haqq-Misra, J. (2011) A protocol for messaging to extraterrestrial intelligence. *Space Policy, 27*(3), 165-169.

125 Cook, J. (2010, December 13) Voyager, An Interstellar Journey. NASA Probe Sees Solar Wind Decline. *NASA.gov*. Retrieved December 21, 2015 from: http://www.nasa.gov/mission_pages/voyager/voyager20101213.html

126 NASA Administrator (2015, March 10) Is Warp Drive Real? *NASA.gov*. Retrieved December 22, 2015 from: http://www.nasa.gov/centers/glenn/technology/warp/warp.html

127 Powell, C. S. (2015, May 28) Where's My Warp Drive? *Discover Magazine*. Retrieved September 14, 2015 from: http://discovermagazine.com/2015/july-aug/31-wheres-my-warp-drive

128 Einstein, A., & Rosen, N. (1935) The particle problem in the general theory of relativity. *Physical Review, 48*(1), 73.

129 HarvardX: SPU30x Super-Earths and Life (2015) EdX online course in Astrobiology through Harvard University's HarvardX program.

130 Thorne, K. S. (1994) *Black Holes & Time Warps: Einstein's Outrageous Legacy* (pp. 483). New York, New York: WW Norton & Company.

131 Thorne, K. (1995) *Black Holes & Time Warps: Einstein's Outrageous Legacy* (pp. 451-452). New York, New York: WW Norton & Company.

132 Thorne, K. S. (1994) *Black Holes & Time Warps: Einstein's Outrageous Legacy*. New York, New York: WW Norton & Company.

133 Thorne, K. S. (1994) *Black Holes & Time Warps: Einstein's Outrageous Legacy* (pp. 128-130). New York, New York: WW Norton & Company.

134 Thorne, K. (1995) *Black Holes & Time Warps: Einstein's Outrageous Legacy* (pp. 493). New York, New York: WW Norton & Company.

135 Vakoch, D. (2014) Reconstructing Distant Civilizations and Encountering Alien Cultures. *Archaeology, Anthropology, and Interstellar Communication* (pp. xiv). Vakoch, D. A. (Ed.). National Aeronautics and Space Administration. Office of Communications, Public Outreach Division.

136 Horowitz, S. S. (2012) *The Universal Sense: How Hearing Shapes the Mind*. Bloomsbury Publishing USA.

137 McQuay B. & Joyce, C. (2015, September 10) How Sound Shaped The Evolution Of Your Brain. *National Public Radio*. Retrieved September 14, 2015 from: http://www.npr.org/sections/health-shots/2015/09/10/436342537/how-sound-shaped-the-evolution-of-your-brain

138 Gardner, R. A., & Gardner, B. T. (1969) Teaching sign language to a chimpanzee. *Science, 165*(3894), 664-672.

139 Douglas Vakoch, describing a chapter and work by anthropologist Ben Finney and historian Jerry Bentley, titled A Tale of Two Analogues - Learning at a Distance from the Ancient Greeks and Maya and the Problem of Deciphering Extraterrestrial Radio Transmissions. *Archaeology, Anthropology, and Interstellar Communication* (2014) (pp. xviii) Vakoch,

D. A. (Ed.). National Aeronautics and Space Administration. Office of Communications, Public Outreach Division.

140 Finney, B. & Bentley, J. (1998) A Tale of Two Analogues - Learning at a Distance from the Ancient Greeks and Maya and the Problem of Deciphering Extraterrestrial Radio Transmissions. *Acta Astronautica* 42, 691-696.

141 NASA Administrator (2007, March 26) The Pioneer Missions. Retrieved July 17, 2018 from: https://www.nasa.gov/centers/ames/missions/archive/pioneer.html

142 Angrum, A. (n.d.) The Golden Record. Voyager, the Interstellar Mission. *Jet Propulsion Laboratory, California Institute of Technology.* Retrieved December 22, 2015 from: http://voyager.jpl.nasa.gov/spacecraft/goldenrec.html

143 Webb, W. (1961) A Dramatic UFO Encounter in the White Mountains, New Hampshire: The Hill Case—Sept. 19-20, 1961. *Confidential NICAP Report, October, 26.* Retrieved October 23, 2015 from: http://www.nicap.org/reports/610919hill_report2.pdf

144 Mead, M., Sieben, A., & Straub, J. (1973). *Coming of age in Samoa.* Penguin.

145 Masetti, M. (2015, July 22) *How Many Stars in the Milky Way?* Blueshift, National Aeronautics and Space Administration, NASA, Astrophysics Science Division. Retrieved December 21, 2015 from: http://asd.gsfc.nasa.gov/blueshift/index.php/2015/07/22/how-many-stars-in-the-milky-way/

146 Howell, E. (2014, April 1) How Many Galaxies Are there? *Space.com.* Retrieved December 22, 2015 from: http://www.space.com/25303-how-many-galaxies-are-in-the-universe.html

Chapter 6

147 Augustine, S. (1876) *The confessions* (Book 11, Chapter 14) Clark.

148 Friedman, J., Morris, M. S., Novikov, I. D., Echeverria, F., Klinkhammer, G., Thorne, K. S., & Yurtsever, U. (1990) Cauchy problem in spacetimes with closed timelike curves. *Physical Review D, 42*(6), 1915.

149 Deutsch, D. (1991) Quantum mechanics near closed timelike lines. *Physical Review D, 44*(10), 3197.

150 Ringbauer, M., Broome, M. A., Myers, C. R., White, A. G., & Ralph, T. C. (2014) Experimental simulation of closed timelike curves. *Nature communications, 5.*

151 Novikov, I. D. (1983) Evolution of the Universe. Cambridge, Cambridge University Press, 190 p. Translation., 1.

152 Carlini, A., Frolov, V. P., Mensky, M. B., Novikov, I. D., & Soleng, H. H. (1995) Time machines: the Principle of Self-Consistency as a consequence of the Principle of Minimal Action. *International Journal of Modern Physics D, 4*(05), 557-580.

153 Hawking, S. W. (1992) Chronology protection conjecture. *Physical Review D, 46*(2), 603.

154 Earman, J., Smeenk, C., & Wüthrich, C. (2009) Do the laws of physics forbid the operation of time machines?. *Synthese, 169*(1), 91-124.

155 Earman, J. (1995) Outlawing time machines: Chronology protection theorems. *Erkenntnis, 42*(2), 125-139.

156 Droit-Volet, S., Brunot, S., & Niedenthal, P. (2004). BRIEF REPORT Perception of the duration of emotional events. *Cognition and Emotion, 18*(6), 849–858.

157 Gable, P. A., & Poole, B. D. (2012). Time flies when you're having approach-motivated fun: Effects of motivational intensity on time perception. *Psychological science, 23*(8), 879–886.

158 Chen, M. K. (2013) The effect of language on economic behavior: Evidence from savings rates, health behaviors, and retirement assets. *The American Economic Review, 103*(2), 690–731.

159 Whorf, B. & Carroll, J. B. (1956) Language, thought, and reality: Selected writings of Benjamin Lee Whorf. MIT Press.

160 Frank, M. C., Everett, D. L., Fedorenko, E., & Gibson, E. (2008) Number as a cognitive technology: Evidence from Pirahã language and cognition. *Cognition, 108*(3), 819-824.

161 Corballis, M. C., & Suddendorf, T. (2007) 2: Memory, Time and Language. *What makes us human?17*, 29.

162 Heynick, F. (1983) From Einstein to Whorf: space, time, matter, and reference frames in physical and linguistic relativity. *Semiotica, 45*(1-2), 35-64.

163 Kay, P., & Kempton, W. (1984) What is the Sapir-Whorf hypothesis?. *American Anthropologist, 86*(1), 65-79.

164 Boroditsky, L. (2001) Does language shape thought?: Mandarin and English speakers' conceptions of time. *Cognitive psychology, 43*(1), 1-22.

165 Levin, I., & Zakay, D. (Eds.). (1989) Time and human cognition: A life-span perspective. Elsevier.

166 Augustine, S. (1876) *The confessions* (Chapter XXVII). Clark.

167 Gleiser, M. (2014) *The island of knowledge: The limits of science and the search for meaning.* Basic Books.

168 From: Davies, P. (1996) About time: Einstein's unfinished revolution. Simon and Schuster. Page 77. Citing a quote by Albert Einstein in The Philosophy of Rudolf Carnap, ed. P.A. Schilpp (Open Court, la Salle, Ill., 1963), pg. 37.

169 Bruss, F. T., & Rüschendorf, L. (2010) On the perception of time. *Gerontology, 56*(4), 361-370.

170 Gonzalez-Bellido, P. T., Fabian, S. T., & Nordström, K. (2016). Target detection in insects: optical, neural and behavioral optimizations. *Current opinion in neurobiology, 41*, 122–128.

171 Galloway, R. (2017 September 17) Why is it so hard to swat a fly? BBC News Science and Environment. Retrieved May 18, 2018 from: http://www.bbc.com/news/science-environment-41284065

172 Wang, H. (1995). Time in philosophy and in physics: From Kant and Einstein to Gödel. *Synthese, 102*(2), 215-234, quote from page 220.

173 Onion, A. (2002 March 6) Scientists Explain Why Time Travel Is Possible. ABC News. Retrieved January, 2016 from: http://abcnews.go.com/Technology/story?id=98062&page=1

174 Øhrstrøm, P., & Hasle, P. F. (1995) Temporal logic: from ancient ideas to artificial intelligence (Vol. 57). *Springer Science & Business Media.*

175 Cowan, H. J. (1958) Time and its measurement; from the stone age to the nuclear age. Cleveland, World Pub. Co.[1958][1st ed.]., 1.

176 United States Navy. Cesium Atoms at Work. Retrieved May, 2015 from: http://tycho.usno.navy.mil/cesium.html

177 Einstein, A. (1905) On the electrodynamics of moving bodies. Annalen der Physik, 17(891), 50.

178 Einstein, A. (1915) Zur allgemeinen Relativitätstheorie. Sitzungsber. Kön. Preuß. Akad. Wiss. zu Berlin, 778–786.

179 Einstein, A. (1915) Zur allgemeinen Relativitätstheorie (Nachtrag). Sitzungsber. Kön. Preuß. Akad. Wiss. zu Berlin, 799–801.

180 Einstein, A. (1915) Die Feldgleichungen der Gravitation. Sitzungsber. Kön. Preuß. Akad. Wiss. Zu Berlin, 844–847.

181 Lobo, F., & Crawford, P. (2003) Time, closed timelike curves and causality. *In The Nature of Time: Geometry, Physics and Perception* (pp. 289-296). Springer Netherlands.

182 Hafele, J. C., & Keating, R. E. (1972) Around-the-world atomic clocks: predicted relativistic time gains. *Science, 177*(4044), 166-168.

183 Ludlow, A. D., Boyd, M. M., Ye, J., Peik, E., & Schmidt, P. O. (2015) Optical atomic clocks. *Reviews of Modern Physics, 87*(2), 637.

184 Chou, C. W., Hume, D. B., Rosenband, T., & Wineland, D. J. (2010) Optical clocks and relativity. *Science, 329*(5999), 1630-1633.

185 Dr. William Phillips, interviewed for ElectronCafe.com. Retrieved November, 2014 from: http://www.electricalfun.com/ElectronCafe/William_Phillips_interview.aspx

186 Davies, P. (1996) About time: Einstein's unfinished revolution. Simon and Schuster.

187 Reynolds, S. (2003) Analyzing Time in Cross-cultural Communication: A different approach. Time: Linear, Flexible or Cyclical. ABC Europe 2003 Conference. Retrieved August 3, 2016 from: http://consulting-success.org/wp/?page_id=1204

188 Kitching, T. (2016 February 25) What Is Time – And Why Does It Move Forward? *The Conversation.* Retrieved February 29, 2016 from: https://the-

conversation.com/what-is-time-and-why-does-it-move-forward-55065

189 Haynie, D. T. (2001) Biological thermodynamics. Cambridge University Press.

190 England, J. L. (2013) Statistical physics of self-replication. *The Journal of chemical physics, 139*(12), 121923.

191 England, J. L. (2015). Dissipative adaptation in driven self-assembly. *Nature nanotechnology, 10*(11), 919-923.

192 Perunov, N., Marsland, R., & England, J. (2016) Statistical physics of adaptation. *Physical Review X.* 6(2)021036(12)

193 Manning, A. G., Khakimov, R. I., Dall, R. G., & Truscott, A. G. (2015) Wheeler's delayed-choice gedanken experiment with a single atom. *Nature Physics.* doi:10.1038/nphys3343.

194 Wheeler, J. A. in *Mathematical Foundations of Quantum Theory* (ed Marlow, A. R.) (Academic Press, 1978).

195 Einstein, A. (1999) Albert Einstein Quotes. Retrieved June, 2015 from: http://www.rarre.org/documents/einstein/Collected Quotes from AlberEinstein-2.pdf

196 Frank, A. (2016 February 16) Was Einstein Wrong? National Public Radio blog 13.7: Cosmos and Culture. Retrieved March 15, 2016 from: http://www.npr.org/sections/13.7/2016/02/16/466109612/was-einstein-wrong?utm_source=facebook.com&utm_medium=social&utm_campaign=npr&utm_term=nprnews&utm_content=20160216

197 Davies, P. (1996) About time: Einstein's unfinished revolution, pg. 76. Simon and Schuster.

198 Libet, B., Gleason, C. A., Wright, E. W., & Pearl, D. K. (1983) Time of conscious intention to act in relation to onset of cerebral activity (readiness-potential). *Brain, 106*(3), 623-642.

199 See for example Haggard, P., Newman, C., & Magno, E. (1999) On the perceived time of voluntary actions. *British Journal of Psychology, 90*(2), 291-303.

200 Fried, I., Mukamel, R., & Kreiman, G. (2011) Internally generated preactivation of single neurons in human medial frontal cortex predicts volition. *Neuron, 69*(3), 548-562.

201 Haggard, P., Clark, S., & Kalogeras, J. (2002) Voluntary action and conscious awareness. *Nature neuroscience, 5*(4), 382-385.

202 Blakemore, S. J., Frith, C. D., & Wolpert, D. M. (1999) Spatio-temporal prediction modulates the perception of self-produced stimuli. *Journal of cognitive neuroscience, 11*(5), 551-559.

203 Blakemore, S. J., Wolpert, D., & Frith, C. (2000) Why can't you tickle yourself?. *Neuroreport, 11*(11), R11-R16.

204 Frith, C. (2013) Making up the mind: How the brain creates our mental world. John Wiley & Sons.

205 Zeldovich, I. B., & Novikov, I. D. (1975) Structure and Evolution of the Universe. Izdatel'stvo Nauka, Moscow.

206 Zel'dovich, I. B., & Novikov, I. D. (1983) The Structure and Evolution of the Universe. University of Chicago Press.

207 Novikov, I. D. (1983) Evolution of the Universe. Cambridge, Cambridge University Press.

208 Echeverria, F., Klinkhammer, G., & Thorne, K. S. (1991) Billiard balls in wormhole spacetimes with closed timelike curves: Classical theory. *Physical Review D, 44*(4), 1077.

209 Thorne, K. S. (1991) Do the Laws of Physics Permit Closed Timelike Curves? a. *Annals of the New York Academy of Sciences, 631*(1), 182-193.

210 Mikheeva, E. V., & Novikov, I. D. (1993) Inelastic billiard ball in a spacetime with a time machine. *Physical Review D,* 47(4), 1432.

211 Novikov, I. D. (1998) The River of Time (pp. 253). Cambridge University Press.

212 Novikov, I. D. (1998) The River of Time. Cambridge University Press.

213 Thorne, K. (1994) Black Holes & Time Warps: Einstein's Outrageous Legacy (Commonwealth Fund Book Program). WW Norton & Company.

214 Deutsch, D., & Lockwood, M. (1994) The quantum physics of time travel. Scientific American, 270, 68-74. Described in Billings, L. (2014 September 2) Time Travel Simulation Resolves "Grandfather Paradox." Scientific American. Retrieved May, 2015 from: http://www.scientificamerican.com/article/time-travel-simulation-resolves-grandfather-paradox/

215 Quoted in Billings, L. (2014 September 2) Time Travel Simulation Resolves "Grandfather Paradox". Scientific American. Retrieved May 22, 2015 from: http://www.scientificamerican.com/article/time-travel-simulation-resolves-grandfather-paradox/

216 Friedman, J., Morris, M. S., Novikov, I. D., Echeverria, F., Klinkhammer, G., Thorne, K. S., & Yurtsever, U. (1990) Cauchy problem in spacetimes with closed timelike curves (pp. 1927). *Physical Review D, 42*(6), 1915.

217 Futurama TV Series (2001) Roswell That Ends Well, episode 19 season 3. Created by Matt Groening and David X. Cohen. Fox Broadcasting Company.

218 Levine, A. G. (2010) The Futurama of Physics with David X. Cohen. American Physics Society, Profiles in Versatility. Retrieved April, 2015 from: http://www.aps.org/publications/apsnews/201005/profiles.cfm.

219 Verrone, P.M. (2014 September 21) Welcome to the War of Tomorrow. How *Futurama*'s writers depicted asymmetrical warfare. *Slate magazine.* Retrieved May, 2015 from: http://www.slate.com/articles/technology/future_tense/2014/09/futurama_writer_patric_verrone_on_how_the_cartoon_depicted_asymmetrical.html

Chapter 7

220 Renowned physicist David Deutsch speaking about the possibility of backward time travel on the PBS NOVA program "Time Travel" Airdate: October 12, 1999. Retrieved May, 2015 from: http://www.pbs.org/wgbh/nova/transcripts/2612time.html

221 Abbott, E.A. (1884) *Flatland. A Romance of Many Dimensions.* Seeley and Co. Ltd., London. (Reprinted in 1992 by Dover Publications).

222 Bonnor, W. B. (2003) Closed timelike curves in classical relativity. *International Journal of Modern Physics D, 12*(09), 1705-1708.

223 Penrose, Roger (2004) *The Road to Reality: A Complete Guide to the Laws of the Universe* (Section 17.7). New York: Vintage Books.

224 Lobo, F., & Crawford, P. (2003) Time, closed timelike curves and causality. *The Nature of Time: Geometry, Physics and Perception* (pp. 289-296). Springer Netherlands.

225 Stein, Leo C. (2010, May 11) *What are Closed Timelike Curves?* Retrieved April, 2015 from: http://www.quora.com/What-are-closed-timelike-curves.

226 Van Stockum, W. J. (1938) IX.—The Gravitational Field of a Distribution of Particles Rotating about an Axis of Symmetry. *Proceedings of the Royal Society of Edinburgh, 57*, 135-154.

227 Pickover, C. A. (1998) *Time: A Traveler's Guide.* Oxford University Press.

228 Pfister, H. (2007) On the history of the so-called Lense-Thirring effect. *General Relativity and Gravitation, 39*(11), 1735-1748.

229 Cui, W., Zhang, S. N., & Chen, W. (1998) Evidence for frame-dragging around spinning black holes in X-ray binaries. *The Astrophysical Journal Letters, 492*(1), L53.

230 Isbell, D. Chandler, L. (1998, March 27) Earth dragging space and time as it rotates. NASA, Release: 98-51. Retrieved August, 2018 from: http://www.nasa.gov/home/hqnews/1998/98-051.txt

231 Ciufolini, I., & Pavlis, E. C. (2004) A confirmation of the general relativistic prediction of the Lense–Thirring effect. *Nature, 431*(7011), 958-960.

232 Ramanujan, K. (2004, October) As World Turns it Drags Time and Space. *Goddard Space Flight Center.* Retrieved August, 2018 from: http://www.nasa.gov/centers/goddard/earthandsun/earth_drag.html

233 Gödel, K. (1949) An example of a new type of cosmological solutions of Einstein›s field equations of gravitation. *Reviews of Modern Physics, 21*(3), 447.

234 Grave, F., Buser, M., Müller, T., Wunner, G., & Schleich, W. P. (2009) The Gödel universe: Exact geometrical optics and analytical investigations on motion. *Physical Review D, 80*(10), 103002.

235 Thorne, K. S. (1992) Closed Timelike Curves in General Relativity and Gravitation. *Proceedings of the 13th International Conference on General Relativity and Gravitation*, edited by R.J. Gleiser, C.N. Kozameh, and O.M.

Moreschi, (Institute of Physics Publishing, Bristol, England, 1993), pp. 295-315.

236 Tipler, F. J. (1974) Rotating cylinders and the possibility of global causality violation. *Physical Review D, 9*(8), 2203.

237 Tipler, F. J. (1974) Rotating cylinders and the possibility of global causality violation. *Physical Review D, 9*(8), 2203. Quote from page 2205.

238 Gribbin, J. (1980) Building A Time Machine. *New Scientist, 87*(1216), 654-660.

239 Hawking, S.W., & Ellis, G.F.R. (1973) *The Large Scale Structure of Spacetime.* Cambridge University Press, London.

240 Harrigan, R. M. (1983) *U.S. Patent No. 4,382,245.* Washington, DC: U.S. Patent and Trademark Office.

241 McClintock, J. E., Shafee, R., Narayan, R., Remillard, R. A., Davis, S. W., & Li, L. X. (2006) The spin of the near-extreme Kerr black hole GRS 1915+ 105. *The Astrophysical Journal, 652*(1), 518.

242 Bousso, R., & Engelhardt, N. (2015) New Area Law in General Relativity. *Physical review letters, 115*(8), 081301.

243 Bonnor, W. B., & Sackfield, A. (1968) The interpretation of some spheroidal metrics. *Communications in Mathematical Physics, 8*(4), 338-344.

244 Morgan, T., & Morgan, L. (1969) The gravitational field of a disk. *Physical Review, 183*(5), 1097.

245 Krogh, F. T., Ng, E. W., & Snyder, W. V. (1982) The gravitational field of a disk. *Celestial mechanics, 26*(4), 395-405.

246 Klein, C., & Richter, O. (1999) Exact relativistic gravitational field of a stationary counterrotating dust disk. *Physical review letters, 83*(15), 2884.

247 Frauendiener, J., & Klein, C. (2001) Exact relativistic treatment of stationary counter rotating dust disks: Physical properties. *Physical Review D, 63*(8), 084025.

248 Ori, A. (2007) Formation of closed timelike curves in a composite vacuum/ dust asymptotically-flat spacetime. *arXiv preprint gr-qc/0701024.*

249 González, G. A., & López-Suspes, F. (2011) Timelike and null equatorial geodesics in the Bonnor-Sackfield relativistic disk. *Revista Integración, 29*(1), 59-72.

250 Füzfa, A. (2016) How current loops and solenoids curve spacetime. *Physical Review D, 93*(2), 024014.

251 Deutsch, D. (1991) Quantum mechanics near closed timelike lines. *Physical Review D, 44*(10), 3197.

252 Brun, T. A. (2003) Computers with closed timelike curves can solve hard problems efficiently. *Foundations of Physics Letters, 16*(3), 245-253.

253 Aaronson, S., & Watrous, J. (2009, February) Closed timelike curves make quantum and classical computing equivalent. In *Proceedings of the Royal Society of London A: Mathematical, Physical and Engineering Sciences* (Vol. 465,

No. 2102, pp. 631-647). The Royal Society.

254 Brun, T. A., & Wilde, M. M. (2012) Perfect state distinguishability and computational speedups with postselected closed timelike curves. *Foundations of Physics, 42*(3), 341-361.

255 The National UFO Reporting Center. Retrieved March, 2015 from: http://www.nwlink.com/~ufocntr/

256 Hawking S. (2016) *Can We Time Travel? Genius.* Produced by Stephen Hawking, the Public Broadcasting Service (PBS), and National Geographic International. Original air date, May 18, 2016. The full-length program can be viewed on PBS at http://www.pbs.org/video/2365757267/ and this particular segment on backward time travel begins at 21:30.

257 Evans, R. (2015). *Greenglow & the search for gravity control.* Troubador Publishing Ltd.

258 Young, N. (2016, March 23) Project Greenglow and the battle with gravity. *BBC News.* Retrieved April 8, 2016 from: http://www.bbc.com/news/magazine-35861334

259 Meek, J. (2000, March 26) BAE's anti-gravity research braves X-Files ridicule. *The Guardian, Science.* Retrieved, August 19, 2016 from: https://www.theguardian.com/science/2000/mar/27/uknews

Chapter 8

260 Johanson, D. (2009, March 6) *Anthropologist Donald Johanson on 'Lucy's Legacy'.* Interviewer: I. Flatow [Transcript]. 90.9 WBUR, Boston's NPR News Station. Retrieved January 28, 2016 from: http://www.wbur.org/npr/101547347

261 Latimer, B. (2005) Editorial: The Perils of Being Bipedal. *Annals of biomedical engineering, 33*(1), 3-6.

262 Centers for Disease Control and Prevention (2015 March 2) CDC presents updated estimates of flu vaccine effectiveness for the 2014–2015 season. Retrieved January 18, 2016 from: http://www.cdc.gov/flu/news/updated-vaccine-effectiveness-2014-15.htm.

263 Marill, M. C. (2015) After flu vaccine mismatch, calls for delayed selection intensify. *Nature medicine, 21*(4), 297-298.

264 U.S. Securities and Exchange Commission (2010, July 28) Mutual Funds, Past Performance. Retrieved January 20, 2016 from: http://www.sec.gov/answers/mperf.htm

265 See Navarrete, A., van Schaik, C. P., & Isler, K. (2011) Energetics and the evolution of human brain size. *Nature, 480*(7375), 91-93.

266 Hublin, J. J., Neubauer, S., & Gunz, P. (2015) Brain ontogeny and life history in Pleistocene hominins. *Philosophical Transactions of the Royal Society of London B: Biological Sciences, 370*(1663), 20140062.

267 Enlow, D. H., & Hans, M. G. (1996) *Essentials of facial growth* (pp. 259-260). Philadelphia: Saunders.

268 Wehr, P. A. (2005) *Three Theories for Facial Paedomorphosis in Human Evolution and the Preference for Facial Underdevelopment* (Doctoral dissertation, California State University, Long Beach).

269 Brace, C. L. (1967) Environment, tooth form, and size in the Pleistocene. *Journal of Dental Research, 46*(5), 809-816.

270 Joyce, C. (2010, August 2) Food for Thought: Meat-Based Diet Made Us Smarter. *National Public Radio.* Retrieved August 22, 2018 from: http://www.npr.org/2010/08/02/128849908/food-for-thought-meat-based-diet-made-us-smarter

271 Zink, K. D., & Lieberman, D. E. (2016) Impact of meat and Lower Palaeolithic food processing techniques on chewing in humans. *Nature.*

272 Main, D. (2013, March 13) Ancient Mutation Explains Missing Wisdom Teeth. *LiveScience.* Retrieved March 24, 2016 from: http://www.livescience.com/27529-missing-wisdom-teeth.html#

273 Stanford, C. B., & Bunn, H. T. (2001) *Meat-eating & human evolution.* Oxford: Oxford University Press.

274 Aiello, L. C., & Wheeler, P. (1995) The expensive-tissue hypothesis: the brain and the digestive system in human and primate evolution. *Current anthropology, 36*(2), 199-221.

275 Tobias, P. V. (1987) The brain of Homo habilis: A new level of organization in cerebral evolution. *Journal of Human Evolution, 16*(7), 741-761.

276 Broadfield, D. C., Holloway, R. L., Mowbray, K., Silvers, A., Yuan, M. S., & Màrquez, S. (2001) Endocast of Sambungmacan 3 (Sm 3): a new Homo erectus from Indonesia. *The Anatomical Record, 262*(4), 369-379.

277 Holloway Jr, R. L. (1969) Culture: a human domain. *Current Anthropology,* 395-412.

278 Toth, N. (1985) Archaeological evidence for preferential right-handedness in the Lower and Middle Pleistocene, and its possible implications. *Journal of Human Evolution, 14*(6), 607-614.

279 Wrangham, R. (2009) *Catching fire: how cooking made us human.* New York, New York: Basic Books.

280 Attwell, L., Kovarovic, K., & Kendal, J. R. (2015) Fire in the Plio-Pleistocene: the functions of hominin fire use, and the mechanistic, developmental and evolutionary consequences. *Journal of Anthropological Sciences, 93*, 1-20.

281 Wrangham, R. W., Jones, J. H., Laden, G., Pilbeam, D., & Conklin-Brittain, N. (1999) The raw and the stolen. *Current anthropology, 40*(5), 567-594.

282 Rowlett, R. M. (2000) Fire control by Homo erectus in East Africa and Asia. *Acta Anthropologica Sinica, 19*(5uppl), 198-208.

283 Wrangham, R. & Carmody, R. (2010) Human adaptation to the control of fire. *Evolutionary Anthropology: Issues, News, and Reviews, 19*(5), 187-199.

284 Katz, D. C., Grote, M. N., & Weaver, T. D. (2017) Changes in human skull morphology across the agricultural transition are consistent with softer diets in preindustrial farming groups. *Proceedings of the National Academy of Sciences, 114*(34), 9050-9055.

285 Clark, J. D. & Harris, J. W. (1985) Fire and its roles in early hominid lifeways. *African Archaeological Review, 3*(1), 3-27.

286 Burton, F. D. (2011) *Fire: The spark that ignited human evolution.* UNM Press.

287 Bickerton, D. (1981) *Roots of language.* Karoma.

288 Bickerton, D. (1992) *Language and species.* University of Chicago Press.

289 Bickerton, D. (2003) Symbol and structure: a comprehensive framework for language evolution. *Studies in the evolution of language, 3*,77-93.

Chapter 9

290 Strieber, W., & McDowall, R. (1987) *Communion: A true story* (pp. 223). Sag Harbor, New York: Beech Tree Books.

291 Carlson DS. (1976) Temporal variation in prehistoric Nubian crania. *Am J Phys Anthropol 45*(3), 467-84.

292 Henneberg, M. (1988) Decrease of human skull size in the Holocene. *Hum Biol, 60*(3), 395-405.

293 Lahr M., Wright R. (1996) The question of robusticity and the relationship between cranial size and shape in *Homo sapiens. J Hum Evol 31*(2), 157-191.

294 Wu, X., Wu L., Zhang Q., Zhu, H., Norton C. (2007) Craniofacial morphological microevolution of Holocene populations in northern China. *Chinese Science Bulletin. 52*(12), 1661-1668.

295 Kidder, J. H., Jantz, R. L. & Smith, F. H. (1992) Defining modern humans: A multivariate approach. In (G. Bräuer & F. H. Smith, Eds) *Continuity or Replacement*, pp. 157–177. Rotterdam: Balkema.

296 Bastir, M., Rosas, A., Gunz, P., Peña-Melian, A., Manzi, G., Harvati, K., & Hublin, J. J. (2011) Evolution of the base of the brain in highly encephalized human species. *Nature communications, 2*, 588.

297 Hublin, J. J., Neubauer, S., & Gunz, P. (2015) Brain ontogeny and life history in Pleistocene hominins. *Philosophical Transactions of the Royal Society of London B: Biological Sciences, 370*(1663), 20140062.

298 Bruner, E., Preuss, T. M., Chen, X., & Rilling, J. K. (2016) Evidence for expansion of the precuneus in human evolution. *Brain Structure and Function, 1*-8.

299 Neubauer, S., Hublin, J. J., & Gunz, P. (2018) The evolution of modern human brain shape. *Science advances, 4*(1), eaao5961.

300 Strieber, W., & McDowall, R. (1987) *Communion: A true story.* Sag Harbor, New York: Beech Tree Books.

301 Strieber, W., & McDowall, R. (1987) *Communion: A true story* (pp. 29). Sag Harbor, New York: Beech Tree Books.

302 Strieber, W., & McDowall, R. (1987) *Communion: A true story* (pp. 68). Sag Harbor, New York: Beech Tree Books.

303 Watson, N. (March 23, 1999) Alien Sex 101, The Antonio Villas Boas Account. Retrieved May, 2015 from: http://www.ufocasebook.com/aliensex101.html

304 Strieber, W., & McDowall, R. (1987) *Communion: A true story* (pp. 246). Sag Harbor, New York: Beech Tree Books.

305 UFO Casebook. Alien Abduction Case Files. Composed and Edited by B J Booth; www.ufocasebook.com. Retrieved February 15, 2016 from: http://www.ufocasebook.com/alienabductions.html

306 The Betty Andreasson Abduction (n.d.) retrieved March 3, 2016 from http://www.ufocasebook.com/Andreasson.html.

307 Booth, B. J. (n.d.) The Buff Ledge Abduction. Retrieved March 3, 2016 from http://www.ufocasebook.com/Buffledge.html. Sources referenced: Webb, W. N. (1994) *Encounter at Buff Ledge: A UFO Case History.*

308 Booth, B. J. (n.d.) Travis Walton Abduction, Part 1 and Part II. Retrieved March 3, 2016 from http://www.ufocasebook.com/Walton.html

309 Abductee Jan Wolski Interviewed-30th Anniversary of Emilcin Abduction (n.d.) retrieved March 3, 2016 from http://www.ufocasebook.com/2008/wolski.html. Sources referenced: www.npn.org.pl

310 1988, DNA Sample from Khoury Abduction Raises Big Questions (n.d.) retrieved March 3, 2016 from http://www.ufocasebook.com/khouryabduction.html. Sources referenced: Chalkley, B. (1999) Strange Evidence. *International UFO Reporter.*

311 1983-The Albert Burtoo Abduction Rejection (n.d.) Retrieved March 3, 2016 from http://www.ufocasebook.com/burtoo.html. Sources referenced: Randles, J. (1997) *Alien contact: the first fifty years* (p. 102). Barnes & Noble.; Good, T. (1996) *Beyond top secret* (pp. 87-93). Sidgwick & Jackson.

312 Abduction on North Canol Road, Northern Canada (2004) retrieved March 3, 2016 from http://www.ufocasebook.com/northcanolroadabduction.html. Source referenced: Jasek, M. (2004) Abduction on the North Canol Road. Retrieved March 3, 2016 from http://www.ufobc.ca/yukon/n-canol-abd/index.htm

313 Allen, B. (2000) The A70 Abduction Case. Sommerville, H. (Ed.). Retrieved March 8, 2016 from http://www.ufocasebook.com/a70abduction.html. Sources referenced: Allen, B. J. & Mott, M. (2010) *Rosslyn, Between Two Worlds.* Healings of Atlantis; Allen, B. J. (2007) *The View from the Abyss.* T. G. S.

314 Harrison, D., Chalker, B. (2001) National Director of The Australian UFO Research Network, Australian Skywatch Director. Retrieved August 28, 2018 from: https://www.ufocasebook.com/gundiahmackay.html. The Gundiah Mackay Abduction Milieu. A preliminary report by Bill Chalker and Diane Harrison AUFORN Director. Retrieved August 28, 2018 from: http://www.auforn.com/Gundiah.html (should this highlighted reference have its own entry?

315 QUFOSR – Australian and Internationals UFO Sightings and Research. The Gundiah Mackay Alien Abduction–Australia. Retrieved August 28, 2018 from: https://qufosr.com/2017/01/09/the-gundiah-mack-ay-alien-abduction-australia/

316 Harrison, D., Chalker, B. (2001) National Director of The Australian UFO Research Network, Australian Skywatch Director. Retrieved August 28, 2018 from: https://www.ufocasebook.com/gundiahmackay.html. The Gundiah Mackay Abduction Milieu. A preliminary report by Bill Chalker and Diane Harrison AUFORN Director. Retrieved August 28, 2018 from: http://www.auforn.com/Gundiah.html.

317 Abduction report from Florida, 2003 (n.d.) retrieved March 3, 2016 from http://www.ufocasebook.com/floridaabduction.html. Source referenced: http://www.profindpages.com/news/2004/07/03/MN187.htm

318 Lewontin, R. C. (1972) The apportionment of human diversity, *Evolutionary Biology 6*, 381-398.

319 Relethford, J. H. (1994) Craniometric variation among modern human populations. *American Journal of Physical Anthropology, 95*(1), 53-62.

320 For an interesting take on some other theoretical results of this trend, see *South Park's Goobacks* episode (season 8, episode 7) Originally aired April 28, 2004 on Comedy Central, Viacom. Retrieved May 15, 2015 from: http://southpark.cc.com/full-episodes/s08e06-goobacks

321 IBIS World (May, 2018) Tanning Salons - US Market Research Report. Retrieved September 5, 2018 from: https://www.ibisworld.com/indus-try-trends/market-research-reports/other-services-except-public-ad-ministration/personal-laundry/tanning-salons.html

322 Hoskins, T. (10 February, 2014) Skin-whitening creams reveal the dark side of the beauty industry. The Guardian. Retrieved March 11, 2016 from: http://www.theguardian.com/sustainable-business/blog/skin-whiten-ing-cream-dark-side-beauty-industry

323 Sanusi, V. (9 March, 2016) This Guy Re-Edited A Magazine Cover To Show Apparent Whitewashing On Black Actors. BuzzFeed News. Retrieved March 11, 2016 from: http://www.buzzfeed.com/victoriasanusi/this-guy-re-edited-a-magazine-cover-to-show-apparent-whitewa#.da5bzYN-jA

324 Darwin, C. (1868). *The variation of animals and plants under domestication* (Vol. 2). O. Judd.

325 Belyaev, D. K., & Trut, L. N. (1989). The convergent nature of incipient forms and the concept of destabilizing selection. *Vavilov's Heritage in Modern Biology*, 155-169.

326 Trut, L., Oskina, I., & Kharlamova, A. (2009). Animal evolution during domestication: the domesticated fox as a model. *Bioessays, 31*(3), 349-360.

327 Wilkins, A. S., Wrangham, R. W., & Fitch, W. T. (2014) The "domestication syndrome" in mammals: a unified explanation based on neural crest cell behavior and genetics. *Genetics, 197*(3), 795-808.

328 Theofanopoulou, C., Gastaldon, S., O'Rourke, T., Samuels, B. D., Messner, A., Martins, P. T., ... & Boeckx, C. (2017) Self-domestication in Homo sapiens: Insights from comparative genomics. *PloS one, 12*(10), e0185306.

329 Schönbeck, Y., Talma, H., van Dommelen, P., Bakker, B., Buitendijk, S. E., HiraSing, R. A., & van Buuren, S. (2012) The world's tallest nation has stopped growing taller: the height of Dutch children from 1955 to 2009. *Pediatric research, 73*(3), 371-377.

330 Bogin, B. (1999) *Patterns of human growth* (Vol. 23). Cambridge University Press.

331 Cameron, N., & Bogin, B. (2012) *Human growth and development.* Academic Press.

332 Steckel, R. H., & Rose, J. C. (2002) *The backbone of history: health and nutrition in the Western Hemisphere* (Vol. 2). Cambridge University Press.

333 Sutikna, T., Tocheri, M. W., Morwood, M. J., Saptomo, E. W., Awe, R. D., Wasisto, S., ... & Storey, M. (2016) Revised stratigraphy and chronology for *Homo floresiensis* at Liang Bua in Indonesia. *Nature, 532*(7599), 366.

334 Brown, P., Sutikna, T., Morwood, M. J., Soejono, R. P., Saptomo, E. W., & Due, R. A. (2004) A new small-bodied hominin from the Late Pleistocene of Flores, Indonesia. *Nature, 431*(7012), 1055-1061.

335 Hershkovitz, I., Kornreich, L., & Laron, Z. (2007) Comparative skeletal features between Homo floresiensis and patients with primary growth hormone insensitivity (Laron Syndrome). *American journal of physical anthropology, 134*(2), 198-208.

336 Jungers, W., Baab, K. (2009) The geometry of hobbits: Homo floresiensis and human evolution. *Significance, 6*(4), 159-164.

337 Van Heteren, A. H. (2008) Homo floresiensis as an island form. *Palarch's Journal of Vertebrate Palaeontology, 5*, 1-19.

338 Antón, S. C. (2003) Natural history of Homo erectus. *American journal of physical anthropology, 122*(S37), 126-170.

339 Brown FH, McDougall I. (1993) Geological setting and age. In: Walker A, Leakey R, editors. *The Nariokotome Homo erectus skeleton.* Cambridge, MA: Harvard University Press. p. 9–20.

340 Walker, A., & Leakey, R. E. (1993) *The Nariokotome homo erectus skeleton.* Harvard University Press.

341 Jones, D., Hemphill, W., Meyers, E. (1973) Stature, weight and other physical characteristics of New South Wales schoolchildren. Part 1: Children aged 5 years and over. Special Report Sydney NWS Department of Health.

342 Clegg, M., & Aiello, L. C. (1999). A comparison of the Nariokotome Homo erectus with juveniles from a modern human population. *American Journal of Physical Anthropology, 110*(1), 81-93.

343 McHenry, H. M. (1991) Femoral lengths and stature in Plio-Pleistocene hominids. *American Journal of Physical Anthropology, 85*(2), 149-158.

Chapter 10

344 Degrasse Tyson, N. (2005, May 2) *Einstein and Darwin: A Tale of Two Theories/ Interviewer: Boyle, A.* [Transcript]. NBC News. Retrieved January 28, 2016 from: http://www.nbcnews.com/id/7159345/ns/technology_and_sci-ence-science/t/einstein-darwin-tale-two-theories/#.Vqp-csU6x65

345 Sagan, C. E. (1980) Episode 2: One voice in the cosmic fugue [Television series episode]. *A. Malone (Producer), Cosmos: A Personal Voyage. Arlington, VA: Public Broadcasting Service.*

346 Warrener, A. G., Lewton, K. L., Pontzer, H., & Lieberman, D. E. (2015) A wider pelvis does not increase locomotor cost in humans, with implications for the evolution of childbirth. *PloS one, 10*(3), e0118903.

347 Dunsworth, H. M., Warrener, A. G., Deacon, T., Ellison, P. T., & Pontzer, H. (2012) Metabolic hypothesis for human altriciality. *Proceedings of the National Academy of Sciences, 109*(38), 15212-15216.

348 Betti, L., & Manica, A. (2018). Human variation in the shape of the birth canal is significant and geographically structured. *Proc. R. Soc. B, 285*(1889), 20181807.

349 Abitbol, M. M. (1996) *Birth and human evolution.* Bergin & Garvey.

350 Keeler, C. E. (2010) Land of the moon-children: the primitive San Blas culture in flux. University of Georgia Press.

351 Weiner, S., Monge, J., & Mann, A. (2008) Bipedalism and parturition: an evolutionary imperative for cesarean delivery?. *Clinics in perinatology, 35*(3), 469-478.

352 Sewell, J. E. (1993) Cesarean section-a brief history. A brochure to accompany an exhibition on the history of cesarean section at the National Library of Medicine, 30.

353 Hamilton, B. E., Martin, J. A., Osterman, M. J., Curtin, S. C., & Matthews, T. J. (2015). Births: Final Data for 2014. National vital statistics reports: from the Centers for Disease Control and Prevention, National Center for Health Statistics, National Vital Statistics System, 64(12), 1-64.

354 For further discussion on this, and other factors relating to the increased frequency of cesarean sections over the last 40 years see Baxter, A. [Interviewer] & Keirns, C. [Interviewee] (2015, January 14) *When to Say No*

to a C-Section: How to talk with your doctor about your delivery options [Interview Transcript]. PBS Newshour. Retrieved April 21, 2016 from: http://www.pbs.org/newshour/updates/when-to-say-no-to-a-c-section/.

355 See also Keirns, C. (2015, January) Watching the Clock: A Mother's Hope for a Natural Birth in a Cesarean Culture. *Health Affairs, 34*(1), 178-182 doi: 10.1377/hlthaff.2014.0563.

356 See also Wolf, N. (2002) *Misconceptions: Truth, lies and the unexpected on the journey to motherhood.* Random House.

357 Plante, L. A. (2006) Public health implications of cesarean on demand. *Obstetrical & gynecological survey, 61*(12), 807-815.

358 Office on Women's Heath, U.S. Department of Health and Human Services (2010, September 27) Womenshealth.gov. Retrieved December 21, 2016 from: https://www.womenshealth.gov/pregnancy/childbirth-beyond/labor-birth.html#f

359 McDaniel, M. A. (2005) Big-brained people are smarter: A meta-analysis of the relationship between in vivo brain volume and intelligence. *Intelligence, 33*(4), 337-346.

360 Haier, R. J., Jung, R. E., Yeo, R. A., Head, K., & Alkire, M. T. (2004) Structural brain variation and general intelligence. *Neuroimage, 23*(1), 425-433.

361 Witelson, S. F., Beresh, H., & Kigar, D. L. (2006) Intelligence and brain size in 100 postmortem brains: sex, lateralization and age factors. *Brain, 129*(2), 386-398.

362 Wickett, J. C., Vernon, P. A., & Lee, D. H. (2000) Relationships between factors of intelligence and brain volume. *Personality and Individual Differences, 29*(6), 1095-1122.

363 Gómez-Robles, A., Hopkins, W. D., & Sherwood, C. C. (2013) Increased morphological asymmetry, evolvability and plasticity in human brain evolution. *Proceedings of the Royal Society B, 280*(1761), 20130575.

364 Gómez-Robles, A., Hopkins, W. D., Schapiro, S. J., & Sherwood, C. C. (2015) Relaxed genetic control of cortical organization in human brains compared with chimpanzees. *Proceedings of the National Academy of Sciences, 112*(48), 14799-14804.

365 Gaser, C., & Schlaug, G. (2003) Brain structures differ between musicians and non-musicians. *The Journal of Neuroscience, 23*(27), 9240-9245.

366 Luders, E., Narr, K. L., Bilder, R. M., Thompson, P. M., Szeszko, P. R., Hamilton, L., & Toga, A. W. (2007) Positive correlations between corpus callosum thickness and intelligence. *Neuroimage, 37*(4), 1457-1464.

367 Luders, E., Narr, K. L., Thompson, P. M., & Toga, A. W. (2009) Neuroanatomical correlates of intelligence. *Intelligence, 37*(2), 156-163.

368 Ross, C. F. (1995) Allometric and functional influences on primate orbit orientation and the origins of the Anthropoidea. *Journal of Human Evolution, 29*(3), 201-227.

369 Lieberman, D. E., Pearson, O. M., & Mowbray, K. M. (2000) Basicranial influence on overall cranial shape. *Journal of Human Evolution, 38*(2), 291-315.

370 Lieberman, D. E., Ross, C. F., & Ravosa, M. J. (2000) The primate cranial base: ontogeny, function, and integration. *American Journal of Physical Anthropology, 113*(s 31), 117-169.

371 Ravosa, M. J., Noble, V. E., Hylander, W. L., Johnson, K. R., & Kowalski, E. M. (2000) Masticatory stress, orbital orientation and the evolution of the primate postorbital bar. *Journal of Human Evolution, 38*(5), 667-693.

372 Masters, M. P. (2012) Relative size of the eye and orbit: an evolutionary and craniofacial constraint model for examining the etiology and disparate incidence of juvenile-onset myopia in humans. *Medical hypotheses, 78*(5), 649-656.

373 Gould, S. J., & Lewontin, R. C. (1979) The spandrels of San Marco and the Panglossian paradigm: a critique of the adaptationist programme. *Proceedings of the Royal Society of London B: Biological Sciences, 205*(1161), 581-598.

374 Schultz, A. H. (1940) The size of the orbit and of the eye in primates. *Am J Phys Anthropol 26*, 389-408.

375 Chau, A., Fung, K., Pak, K., & Yap, M. (2004) Is eye size related to orbit size in human subjects? *Ophthal Physl Opt 24*, 35-40.

376 Todd, T., Beecher, H., Williams, G., & Todd, A. (1940) The weight and growth of the human eyeball. *Hum Biol ,12*, 1-20.

377 Weale, R. (1982) *A Biography of the Eye: Development, Growth, Age.* H.K. Lewis & Co.

378 Weiss, K. (2002) How the Eye Got its Brain. *Evol Anthr, 11*, 215-219.

379 Miller, E. M. (1992) On the correlation of myopia and intelligence. *Genet Soc Gen Psychol Monogr, 118*, 361-383.

380 Mak, M., Kwan, T., Cheng, K., Chan, R., & Ho, S. (2006) Myopia as a latent phenotype of a pleiotropic gene positively selected for facilitating neurocognitive development, and the effects of environmental factors in its expression. *Med Hypotheses, 66*, 1209-1215.

381 Cheverud, J. (1996) Developmental integration and the evolution of pleiotropy. *American Zoology, 36*, 44-50.

382 Collins, P. (1995) Embryology and development. *Gray's anatomy, 38th edn. Churchill Livingstone, London*, 91-341.

383 Zadnik, Z., Satariano, W. A., Mutti, D. O., Sholtz, R. I., & Adams, A. J. (1994) The effect of parental history of myopia on children's eye size. *J Am Med Assoc 271*, 1323-1327.

384 Ip, J. M., Huynh, S. C., Kifley, A., Rose, K. A., Morgan, I. G., Varma, R., & Mitchell, P. (2007) Variation of the contribution from axial length and other oculometric parameters to refraction by age and ethnicity. *Invest Ophthalmol Vis Sci* 48:4846-4853.

385 Lam, D. S., Fan, D. S., Lam, R. F., Rao, S. K., Chong, K. S., Lau, J. T., Lai, R. Y., & Cheung, E. Y. (2008) The effect of parental history of myopia on children's eye size and growth: results of a longitudinal study. *Invest Ophthalmol Vis Sci, 49,* 873–876.

386 Goldschmidt, E., Lam, C., & Opper, S. (2001) The development of myopia in Hong Kong children. *Acta Ophthalmologica Scandinavica, 79,* 228-232.

387 Park, D. J. & Congdon, N. G. (2004) Evidence for an "epidemic" of myopia. Ann Acad Med Singapore. 33, 21–6.

388 Ip, J. M., Huynh, S. C., Kifley, A., Rose, K. A., Morgan, I., Wang, J., & Mitchell, P. (2008) Ethnic differences in refraction and ocular biometry in a population-based sample of 11-15 year-old Australian children. *Eye, 22,* 649-656.

389 Angle, J. & Wissmann, D. (1980) The Epidemiology of Myopia. *Am J Epidemiol, 11,* 220-228.

390 Grosvenor, T.P. & Goss, D. A. (1990) Clinical Management of Myopia. Boston: Butterworth-Heineman.

391 Parssinen, O. & Lyyra, A. (1993) Myopia and myopic progression among schoolchildren: a three-year follow-up study. *Invest Ophthalmol Vis Sci, 34,* 2794-2802.

392 Cordain, L., Eaton, S. B., Brand Miller, J., Lindeberg, S., & Jensen, C. (2002) An evolutionary analysis of the aetiology and pathogenesis of juvenile-onset myopia. *Acta Ophthalmologica Scandinavica, 80*(2), 125-135.

393 Boyle, A. [Interviewer] & Degrasse Tyson, N. [Interviewee] (2005, May 2) *Einstein and Darwin: A Tale of Two Theories* NBC News. Retrieved January 28, 2016 from: http://www.nbcnews.com/id/7159345/ns/technology_and_science-science/t/einstein-darwin-tale-two-theories/#.VqpcsU6x65

394 McGough, T. Kiviat, R., Seligson, T., Greene, P., Santilli, R. (1995) Alien Autopsy (Fact of Fiction). Released 28 August, 1995. Kiviat/Greene Productions Inc.

395 *Eamonn Investigates: Alien Autopsy*, British Sky Broadcasting. Originally aired on Sky One, 4 April 2006. Transcript retrieved February 2, 2016 from: http://www.outtahear.com/beyond_updates/Alien%20Autopsy%20Proved%20Fake/Transcript%20of%20the%20show%20From%20the%20Eamonn%20Investigates.htm

396 Abduction on North Canol Road, Northern Canada (2004) retrieved March 3, 2016 from http://www.ufocasebook.com/northcanolroadabduction.html. Source referenced: Jasek, M. (2004) Abduction on the North Canol Road. Retrieved March 3, 2016 from http://www.ufobc.ca/yukon/n-canol-abd/index.htm

397 Strieber, W., & McDowall, R. (1987) *Communion: A true story* (pp. 68). Sag Harbor, New York: Beech Tree Books.

398 1988, DNA Sample From Khoury Abduction Raises Big Questions (n.d.)

retrieved March 3, 2016 from http://www.ufocasebook.com/khouryab-duction.html. Sources referenced: Chalkley, B. (1999) Strange Evidence. *International UFO Reporter*.

399 Abductee Jan Wolski Interviewed-30th Anniversary of Emilcin Abduc-tion (n.d.) retrieved March 3, 2016 from http://www.ufocasebook.com/2008/wolski.html. Sources referenced: www.npn.org.pl

400 Murphy, C. J., Zadnik, K., & Mannis, M.J. (1992) Myopia and refractive error in dogs. *Invest. Ophthalmol. Vis. Sci., 33*, 2459-2463.

401 Kubai, M. A., Bentley, E., Miller, P., Mutti, D., & Murphy, C. (2008) Re-fractive states of eyes and association between ametropia and breed in dogs. *Am. J. Vet. Res., 69*, 946-951.

402 Williams, L., Kubai, M., Murphy, C., & Mutti, D. (2011) Ocular Com-ponents in Three Breeds of Dogs with High Prevalence of Myopia. *Optometry Vision Sci., 88*, 269-274.

403 Lorenz, K. (1943) Die angeborenen Formen möglicher erfahrung. *Z. Tierpsy-chol., 5*, 235–409.

404 Golle, J., Lisibach, S., Mast, F. W., & Lobmaier, J. S. (2013) Sweet puppies and cute babies: Perceptual adaptation to babyfacedness transfers across species (pp.4). *PloS one, 8*(3), e58248.

405 For a comprehensive review of the literature on this subject see Luo, L., Ma, X., Zheng, X., Zhao, W., Xu, L., Becker, B., & Kendrick, K. M. (2015). Neural systems and hormones mediating attraction to infant and child faces. *Frontiers in psychology, 6*.

406 Sapolsky, R. (2015, April 8) Why We Melt at Puppy Pictures. *The Wall Street Journal*. Retrieved May, 2015 from: http://www.wsj.com/articles/why-we-melt-at-puppy-pictures-1428504897

407 Beck, A.M., & A.H. Katcher. (1996) *Between pets and people: The importance of animal companionship*. Purdue Univ. Press, West Lafayette, IN.

408 Beck, A. M. (2014) The biology of the human–animal bond. *Animal Frontiers, 4*(3), 32-36.

409 Gould, S. J. (1979) Mickey Mouse meets Konrad Lorenz. *Nat. Hist., 88*(5), 30–36.

410 Poitras, G. (2000). Anime essentials: Everything a fan needs to know.

411 Brooks, R. (2014, January 24) Why the masculine face? Genetic evidence reveals drawbacks of hyper-masculine features. The Conversation. Retrieved September 19, 2018 from: http://theconversation.com/why-the-masculine-face-genetic-evidence-reveals-drawbacks-of-hy-per-masculine-features-22388

412 Perrett, D. I., Lee, K. J., Penton-Voak, I., Rowland, D., Yoshikawa, S., Burt, D. M., & Akamatsu, S. (1998) Effects of sexual dimorphism on facial attractiveness. *Nature, 394*(6696), 884.

413 Gibbons, A. (2014) How we tamed ourselves—and became modern. *Science, 346*(6208), 405-406.

414 Theofanopoulou, C., Gastaldon, S., O'Rourke, T., Samuels, B. D., Messner, A., Martins, P. T., ... & Boeckx, C. (2017). Self-domestication in Homo sapiens: Insights from comparative genomics. *PloS one, 12*(10), e0185306.

415 Hare, B. (2017). Survival of the friendliest: Homo sapiens evolved via selection for prosociality. *Annual review of psychology, 68*, 155-186.

416 Cieri, R. L., Churchill, S. E., Franciscus, R. G., Tan, J., & Hare, B. (2014) Craniofacial feminization, social tolerance, and the origins of behavioral modernity. *Current Anthropology, 55*(4), 419-443.

417 Wilkins, A. S., Wrangham, R. W., & Fitch, W. T. (2014) The "domestication syndrome" in mammals: A unified explanation based on neural crest cell behavior and genetics. *Genetics, 197*(3), 795-808.

418 Thompson, D. W. (1917) *On growth and form.* Cambridge University Press.

419 Cardini, A., & Loy, A. (2013) On growth and form in the" computer era": from geometric to biological morphometrics. In Virtual Morphology and Evolutionary Morphometrics in the New Millennium, eds. Andrea Cardini and Anna Loy. *Hystrix, the Italian Journal of Mammalogy, 24*(1), 1-5.

420 European Virtual Anthropology Network (EVAN) Toolbox software, version 1.70. Developed by EVAN and the EVAN-Society to help facilitate form and shape analysis using Geometric Morphometrics (GM), which includes Thin-Plate Spline Warping, Partial Least Squares Analysis, Principal Component Analysis, and General Procrustes Analysis. Retrieved from: http://www.evan-society.org/node/23

421 Lamm, Nickolay (2013, June 7) What Will Humans Look Like in 100,000 Years? The Code Word. Retrieved May, 2015 from: http://www.myvouchercodes.co.uk/the-code-word/what-will-humans-look-like-in-100000-years/

422 Herper, Matthew (2013, June 7) No, This Is Not How The Human Face Might Look in 100,000 years. Forbes. Retrieved May, 2015 from: http://www.forbes.com/sites/matthewherper/2013/06/07/no-this-is-not-how-the-human-face-might-look-in-100000-years/#334bdfa511f2

Chapter 11

423 Suzuki, D. (2007) *The Sacred Balance: Rediscovering Our Place in Nature, Updated and Expanded.* Greystone Books Ltd.

424 Bohn, L. (2009, March 4) Q&A: 'Lucy' Discoverer Donald C. Johanson. Time. Retrieved May 18, 2016 from: http://content.time.com/time/health/article/0,8599,1882969,00.html

425 Green, R. E., Krause, J., Briggs, A. W., Maricic, T., Stenzel, U., Kircher, M., & Hansen, N. F. (2010) A draft sequence of the Neandertal genome. *Science, 328*(5979), 710-722.

426 Prüfer, K., Racimo, F., Patterson, N., Jay, F., Sankararaman, S., Sawyer, S., & Li, H. (2014). The complete genome sequence of a Neanderthal from the Altai Mountains. *Nature, 505*(7481), 43.

427 Sankararaman, S., Mallick, S., Patterson, N., & Reich, D. (2016). The combined landscape of Denisovan and Neanderthal ancestry in present-day humans. *Current Biology, 26*(9), 1241-1247.

428 Sankararaman, S., Mallick, S., Dannemann, M., Prüfer, K., Kelso, J., Pääbo, S., & Reich, D. (2014) The genomic landscape of Neanderthal ancestry in present-day humans. *Nature, 507*(7492), 354-357.

429 Reich, D., Green, R. E., Kircher, M., Krause, J., Patterson, N., Durand, E. Y., & Maricic, T. (2010) Genetic history of an archaic hominin group from Denisova Cave in Siberia. *Nature, 468*(7327), 1053-1060.

430 Sankararaman, S., Mallick, S., Patterson, N., Reich, D. (2016) The Combined Landscape of Denisovan and Neanderthal Ancestry in Present-Day Humans. *Current Biology.* In Press, doi:10.1016/j.cub.2016.03.037

431 Lee, D. (2006 January 16) *Personal Story: Abducted by Aliens/ Interviewer: Bill O'Reilly* [Transcript]. From: Back of the Book [Television Series Episode]. The O'Reilly Factor: Fox News. Retrieved on February 18, 2016 from: http://www.foxnews.com/story/2006/01/16/personal-story-abducted-by-aliens.html

432 Alford, J. (2014 June 25) Automatic Sperm Extractor Introduced Into A Chinese Hospital. *IFL Science, Technology.* Retrieved May, 2015 from: http://www.iflscience.com/technology/automatic-sperm-extractor-introduced-chinese-hospital.

433 Watson, N. (1999 March 23) Alien Sex 101, The Antonio Villas Boas Account. Retrieved May, 2015 from: http://www.ufocasebook.com/aliensex101.html

434 Strieber, W., & McDowall, R. (1987) *Communion: A true story* (pp. 223). Sag Harbor, New York: Beech Tree Books.

435 Randles, J., Pritchard A., Pritchard D., Mack J., Kasey P., Yapp C (1994) Why are They Doing This? *Alien Discussions: Proceedings of the Abduction Study Conference.* Cambridge: North Cambridge Press. pp. 69–70.

436 Geggel, L. (2016 February 26) Great Dane to Chihuahua: How Do We Know Dogs Are the Same Species? *Livescience.* Retrieved April 4, 2016 from: http://www.livescience.com/53841-how-know-dogs-are-same-species.html

437 Lewontin, R. C. (1972) The apportionment of human diversity, *Evolutionary Biology, 6,* 381-398.

438 Relethford, J. H. (1994) Craniometric variation among modern human populations. *American Journal of Physical Anthropology, 95*(1), 53-62.

439 Ellstrand, N. C., & Elam, D. R. (1993) Population genetic consequences of small population size: implications for plant conservation. *Annual review*

of *Ecology and Systematics*, 217-242.

440 Wright, S. (1949) The genetical structure of populations. *Annals of eugenics*, *15*(1), 323-354.

441 Nei, M. (1973) Analysis of gene diversity in subdivided populations. *Proceedings of the National Academy of Sciences*, *70*(12), 3321-3323.

442 Nei, M. (1977) F-statistics and analysis of gene diversity in subdivided populations. *Annals of human genetics*, *41*(2), 225-233.

443 Wright, S. (1949) The genetical structure of populations. *Annals of eugenics*, *15*(1), 323-354.

444 Wright, S. (1931) Evolution in Mendelian populations. *Genetics*, *16*(2), 97.

445 Holsinger, K. E., & Weir, B. S. (2009) Genetics in geographically structured populations: defining, estimating and interpreting F ST. *Nature Reviews Genetics*, *10*(9), 639.

446 Jim Penniston Hypnosis (1994, September 10) Retrieved September 18, 2018 from: http://www.therendleshamforestincident.com/Penniston_Hypnosis.html

447 Zeng, T. C., Aw, A. J., & Feldman, M. W. (2018) Cultural hitchhiking and competition between patrilineal kin groups explain the post-Neolithic Y-chromosome bottleneck. *Nature communications*, *9*(1), 2077.

448 Zhou, V. (2016, October 27) China has world's most skewed sex ratio at birth–again. South China Morning Post. Retrieved September 26, 2018 from: https://www.scmp.com/news/china/policies-politics/article/2040544/chinas-demographic-time-bomb-still-ticking-worlds-most

449 Abduction on North Canol Road, Northern Canada (2004) retrieved March 3, 2016 from http://www.ufocasebook.com/northcanolroadabduction.html. Source referenced: Jasek, M. (2004) Abduction on the North Canol Road. Retrieved March 3, 2016 from: http://www.ufobc.ca/yukon/n-canol-abd/index.htm

450 Abduction report from Florida, 2003 (n.d.) retrieved March 3, 2016 from http://www.ufocasebook.com/floridaabduction.html. Source referenced: http://www.profindpages.com/news/2004/07/03/MN187.htm

451 Borgenicht, D., & Piven, J. (2012) *The Complete Worst-Case Scenario Survival Handbook*. Chronicle Books. pp. 172.

452 Yoo, S. S., Kim, H., Filandrianos, E., Taghados, S. J., & Park, S. (2013) Non-invasive brain-to-brain interface (BBI): establishing functional links between two brains. *PloS one*, *8*(4), e60410.

453 Rao, R. P., Stocco, A., Bryan, M., Sarma, D., Youngquist, T. M., Wu, J., & Prat, C. S. (2014) A direct brain-to-brain interface in humans. *PloS one*, *9*(11), e111332.

454 BBC News – Technology (2015) Surge in US 'brain-reading' patents. Retrieved May, 2015 from: http://www.bbc.com/news/technology-32623063

455 See HBO's Last Week Tonight with John Oliver for an insightful description of the current problem of patent trolls. Link available through the Business Insider – Tech site: Tweedie S. (April 20th, 2015) Watch John Oliver make fun of patent trolls on 'Last Week Tonight.' Retrieved May, 2015 from: http://www.businessinsider.com/john-oliver-patent-trolls-last-week-tonight-2015-4

456 Nishimura, T., Mikami, A., Suzuki, J., & Matsuzawa, T. (2003) Descent of the larynx in chimpanzee infants. *Proceedings of the National Academy of Sciences, 100*(12), 6930-6933.

457 Lieberman, P. (1968) Primate vocalizations and human linguistic ability. *The Journal of the Acoustical Society of America, 44*(6), 1574-1584.

458 Lieberman, P., Klatt, D. H., & Wilson, W. H. (1969) Vocal tract limitations on the vowel repertoires of rhesus monkey and other nonhuman primates. *Science, 164*(3884), 1185-1187.

459 McCarthy RC, Lieberman DE. (2001) Posterior maxillary (PM) plane and anterior cranial architecture in primates. *Anat Rec, 264*(3), 247-260.

460 Lieberman D., McBratney BM, Krovitz G. (2002) The evolution and development of cranial form in Homo sapiens. *Proc Natl Acad Sci USA 99*(3), 1134-1139.

461 Bastir, M., Rosas, A., Lieberman, D., O'Higgins, P. (2008) Middle Cranial Fossa Anatomy and the Origin of Modern Humans. *The Anatomical Record: Advances in Integrative Anatomy and Evolutionary Biology, 291*(2), 130-140.

Chapter 12

462 Vonnegut, K. (1969) Slaughterhouse Five, or the Children's Crusade: A Duty-Dance with Death. (4.21.5-6) Delacorte. ISBN 0-385-31208-3.

463 Lobo, F., & Crawford, P. (2003) Time, closed timelike curves and causality. In *The Nature of Time: Geometry, Physics and Perception* (pp. 289-296). Springer Netherlands.

464 Deutsch, D. (1991) Quantum mechanics near closed timelike lines. *Physical Review D, 44*(10), 3197.

465 Novikov, I. D. (1998) *The River of Time*. Location: Cambridge University Press.

466 Vonnegut, K. (1969) *Slaughterhouse Five, or the Children's Crusade: A Duty-Dance with Death*. Delacorte. ISBN 0-385-31208-3.

467 Jones, E. M. (1985) "Where is everybody?" An account of Fermi's question. Los Alamos National Laboratory (LANL). United States Department of Energy.

468 Hawking, S., & Jackson, M. (1993) *A brief history of time*. Dove Audio.

469 Davies, P. (1996) *About time: Einstein's unfinished revolution* (pp. 250-251). Simon and Schuster.

470 Davies, P. (1996) *About time: Einstein's unfinished revolution* (pp. 251). Simon and Schuster.

471 Novikov, I. D. (1983) *Evolution of the Universe*. Cambridge, Cambridge University Press, 1983, 190 p. Translation., 1.

472 Friedman, J., Morris, M. S., Novikov, I. D., Echeverria, F., Klinkhammer, G., Thorne, K. S., & Yurtsever, U. (1990) Cauchy problem in spacetimes with closed timelike curves. *Physical Review D, 42*(6), 1915.

473 Novikov, I. D. (1992) Time machine and self-consistent evolution in problems with self-interaction. *Physical Review D, 45*(6), 1989.

474 Deutsch, D., & Lockwood, M. (1994) The quantum physics of time travel. *Scientific American, 270*, 68-74.

475 Lloyd, S., Maccone, L., Garcia-Patron, R., Giovannetti, V., Shikano, Y., Pirandola, S., & Steinberg, A. M. (2011) Closed timelike curves via postselection: theory and experimental test of consistency. *Physical review letters, 106*(4), 040403.

476 Novikov, I. D., & Frolov, V. A. L. E. R. Y. P. (2013) *Physics of black holes* (Vol. 27). Springer Science & Business Media.

477 Ringbauer, M., Broome, M. A., Myers, C. R., White, A. G., & Ralph, T. C. (2014) Experimental simulation of closed timelike curves. *Nature communications, 5.*

478 Billings L. (2014, September 2) Time Travel Simulation Resolves "Grandfather Paradox." *Scientific American.* Retrieved July 18, 2016 from: http://www.scientificamerican.com/article/time-travel-simulation-resolves-grandfather-paradox/

479 For an enlightening insight into the divergent views of these leading researchers of backward time travel, see: Hawking, S. W., Thorne, K. S., Novikov, I. D., Ferris, T., & Lightman, A. (2003) *The future of spacetime.* WW Norton & Company.

480 As an example of Hawking's overt bias against backward time travel, see the mere 10-minute-long segment of a 55-minute episode specifically titled Can We Time Travel? as part of his PBS television series Genius. Hawking S. (2016) Can We Time Travel? Genius. Produced by Stephen Hawking, the Public Broadcasting Service (PBS), and National Geographic International. Original air date, May 18, 2016.

481 Carroll, S. (2010, February 2) The Real Rules for Time Travelers. *Discover Magazine.* Retrieved June 30, 2016 from: http://discovermagazine.com/2010/mar/02-the-real-rules-for-time-travelers

482 Schmalbruch, S. (2015, June 2) The 25 most popular tourist attractions in the world. *Business Insider.* Retrieved January 5, 2018 from: http://www.businessinsider.com/worlds-most-popular-tourist-attractions-2015-6/#25-panama-canal-panama-1

Index

P

Image Credits

Chapter 1

- **Figure 1-1.** A mental image envisioned at age eight, which brought to mind the question of whether "aliens" could simply be humans in the future coming back through time to visit and study us in their own past.

  » Mad Dog/shutterstock.com (left), Makc/shutterstock.com (center), Brian Goff/shutterstock.com (right)

Chapter 3

- **Figure 3-1.** Image of the 2,200 year old *Saqqara Bird* from Saqqara Egypt

  » Dawoud Khalil Messiha [Public domain], via Wikimedia Commons. Retrieved September 23, 2015 from: https://commons.wikimedia.org/wiki/File%3APhoto_1-plane_front_view1.jpg

- **Figure 3-2.** Image of the "Astronaut" geoglyph at Nazca, Peru as viewed from high above.

  » Diego Delso [CC BY-SA 4.0 (http://creativecommons.org/licenses/by-sa/4.0)], via Wikimedia Commons. Retrieved and changes made October 3, 2016 from: https://commons.wikimedia.org/wiki/File%3AL%C3%ADneas_de_Nazca%2C_Nazca%2C_Per%C3%BA%2C_2015-07-29%2C_DD_46.JPG

- **Figure 3-3.** Example of a form of head binding associated with intentional cranial modification, which was most common among the Maya.

  » Fruitpunchline (talk) [Public domain], via Wikimedia Commons. Retrieved January 10, 2019 from: https://commons.wikimedia.org/wiki/File:Maya_cranial_deformation.gif

- **Figure 3-4.** Example of the result of this process observable in the craniofacial anatomy of an adult human dry skull

  » Anna Jurkovska/shutterstock.com

- **Figure 3-5**. Map showing major geographic regions where intentional cranial modification was practiced.

- **Figure 7-2.** A closed timelike curve resulting from light cones tipping over, which may allow time travel into the past.

  » Image created by the author

- **Figure 7-3.** Example of a Levitron hovering above its base by way of magnetic and rotational forces.

  » Netscott, Scott Stevenson. [Public domain], via Wikimedia Commons. Retrieved May 29, 2015 from: https://en.wikipedia.org/wiki/File:Levitron-levitating-top-demonstrating-Roy-M-Harrigans-spin-stabilized-magnetic-levitation.ogg

- **Table 7-1.** Showing some of the 229 sightings reported to NUFORC between 1-1-2014, and 1-1-2015 involving the observed appearance/disappearance of an IFO, which may indicate their passage into, out of, or through the observer's localized region of four-dimensional spactime.

  » Data obtained from the National UFO Reporting Center (NUFORC) database. Monthly Report Index for 01/14 – 1/15. Retrieved May 5, 2015 from: http://www.nuforc.org/

Chapter 8

- **Figure 8-1.** Image showing the effects of basicranial flexing on neurocranial and lower facial shape/size in hominin evolution.

  » Image created by the author

- **Figure 8-2**. Example of the generalized pattern of craniofacial architecture across three different breeds of domesticated dogs. In which smaller and "cuter" breeds possess shorter faces with more bulbous projecting foreheads (right), compared to other breeds with flat foreheads and longer, more projecting faces (left).

  » Image created by the author

- **Figure 8-3.** Showing changes in human body proportions associated with neoteny.

  » From the American Genetic Association (Journal of Heredity (1921) Volume 12, pg 421) [Public domain], via Wikimedia Commons. Retrieved July 15, 2014 from: https://en.wikipedia.org/wiki/File:Neoteny_body_proportion_heterochrony_human.png

- **Figure 8-4.** Illustrating heterochrony, and specifically neoteny in human evolution, which is apparent from our greater resemblance to a juvenile than an adult bonobo chimpanzee.

  » LeonP/shutterstock.com (left), Eric Isselee/shutterstock.com (right)

- **Figure 8-5**. With the continuation of paedomorphic trends into the human future, certain adult characteristics of our extra-tempestrial descendants (exp. proportionally large neurocrania, large eyes, and small faces) may be reflected in the skulls of modern infants.

  » Photo taken by the author

- **Figure 8-6**. Artist depiction of an Upper Paleolithic scene in which lively oral discourse transpires near the light and warmth of a nocturnal fire.

  » Esteban De Armas/shutterstock.com

Chapter 9

- **Figure 9-1.** Artist's depiction of the archetypal physical form of extratempestrials as ubiquitously described in sightings and abduction reports.

  » Bob Orsillo/Shutterstock.com

Chapter 10

- **Figures 10-1 and 10-2.** Images from an alleged autopsy performed on an extratempestrial recovered from the IFO crash at Roswell, New Mexico in 1947. Depicted in a reproduction of the original autopsy film in the documentary *Alien Autopsy: (Fact or Fiction?)*. Originally aired 28 August 1995 (USA) on FOX.

  » Used with permission of Ray Santilli as part of this scientific work.

- **Figure 10-3.** Common modern human cartoon female form showing extreme craniofacial paedomorphosis

  » K.StockPhotos/shutterstock.com

- **Figure 10-4.** Scatterplot with regression line showing a notable change in the eye/head ratio of cartoon characters across an 84-year timespan, signifying the increased cultural value placed on the retention of juvenilized traits since 1932, and most notably since the 1990s. (.emf)

» Graph generated by the author from original data collected as part of this research

- **Figure 10-5.** Anatomical landmarks used in this geometric morphometric analysis to show inter-sex craniofacial shape variability.

  » Image generated using the European Virtual Anthropology Network (EVAN) Toolbox software, version 1.70.

- **Figure 10-6.** Series of images demonstrating shape variation between hypermasculine (left) and hyperfeminine (right) craniofacial forms.

  » Images generated using the European Virtual Anthropology Network (EVAN) Toolbox software, version 1.70.

- **Figure 10-7.** Series of images in the lateral view showing hypermasculine (left) and hyperfeminine (right) craniofacial shape variation.

  » Images generated using the European Virtual Anthropology Network (EVAN) Toolbox software, version 1.70.

Thank you for purchasing and reading *Identified Flying Objects*. If you enjoyed this book, please spread the word and be sure to leave us a review on Amazon.

www.ingramcontent.com/pod-product-compliance
Lightning Source LLC
Chambersburg PA
CBHW060311030426
42336CB00011B/992